EARTH SCIENCES IN THE 21ST CENTURY

MARCELLUS SHALE AND SHALE GAS: FACTS AND CONSIDERATIONS

EARTH SCIENCES IN THE 21ST CENTURY

Additional books in this series can be found on Nova's website under the Series tab.

Additional E-books in this series can be found on Nova's website under the E-books tab.

GASES - CHARACTERISTICS, TYPES AND PROPERTIES

Additional books in this series can be found on Nova's website under the Series tab.

Additional E-books in this series can be found on Nova's website under the E-books tab.

EARTH SCIENCES IN THE 21ST CENTURY

MARCELLUS SHALE AND SHALE GAS: FACTS AND CONSIDERATIONS

GABRIEL L. NAVARRO
EDITOR

Nova Science Publishers, Inc.
New York

For permission to use material from this book please contact us:
Telephone 631-231-7269; Fax 631-231-8175
Web Site: http://www.novapublishers.com

Additional color graphics may be available in the e-book version of this book.

Library of Congress Cataloging-in-Publication Data

Marcellus shale and shale gas : facts and considerations / editor, Gabriel L. Navarro.
p. cm.
Includes bibliographical references and index.
ISBN 978-1-61470-173-6 (hardcover)
1. Gas fields--New York (State)--Marcellus Region. 2. Marcellus Shale. I. Navarro, Gabriel L.
TN872.N7M37 2011
622'.3381--dc23

2011020776

Published by Nova Science Publishers, Inc. † New York

CONTENTS

PREFACE

The Marcellus Shale is a sedimentary rock formation deposited over 350 million years ago in a shallow inland sea located in the eastern United States where the present-day Appalachian Mountains now stand. This shale contains significant quantities of natural gas. New developments in drilling technology, along with higher wellhead prices, have made the Marcellus Shale an important natural gas resource. This book explores water resources and natural gas production from the Marcellus Shale; the impact of Marcellus Shale gas play on current and future Carbon Capture and Storage (CCS) activities and applying this technology to solve America's energy challenges.

Chapter 1- Natural gas is an abundant, domestic energy resource that burns cleanly, and emits the lowest amount of carbon dioxide per calorie of any fossil fuel. The Marcellus Shale and other natural gas resources in the United States are important components of a national energy program that seeks both greater energy independence and greener sources of energy. Marcellus gas development has begun in the northern Appalachian Basin, with significant lease holdings throughout Pennsylvania, West Virginia, southern New York, western Maryland, and eastern Ohio. Because of questions related to water supply and wastewater disposal, however, many state agencies have been cautious about granting permits, and some states have placed moratoriums on drilling until these issues are resolved. At the same time, gas companies, drillers, and landowners are eager to move forward and develop the resource.

Chapter 2- Natural gas represents an important energy source for the United States. According to the U.S. Department of Energy's (DOE's) Energy Information Administration (EIA), about 22% of the country's energy needs are provided by natural gas. Historically, natural gas was produced from conventional vertical wells drilled into porous hydrocarbon-containing formations. During the past decade, operators have increasingly looked to other unconventional sources of natural gas, such as coal bed methane, tight gas sands, and gas shales.

Chapter 3- The Marcellus Shale is an organic rich rock that has been estimated to contain from 50 to 500 trillion cubic feet of natural gas[1]. It was deposited in the Appalachian Basin 350 million years ago as part of an ancient river delta and consists of the bottom layer of an Upper Devonian age sedimentary rock sequence. Like most shale, the Marcellus was deposited as extremely fine grained sediment, with small pore spaces and low permeability that prevents gas from easily migrating[1]. Often called the Marcellus Black Shale due to its color, the formation exists under much of southern New York, Pennsylvania, West Virginia, eastern Ohio, and far western Maryland. Although the shale outcrops at its namesake, Marcellus, New York, it generally lies at depths of 5,000 to 9,000 feet throughout much of the area.[2] The Marcellus Shale generally ranges in thickness from 50 to 200 feet.

Chapter 4- The Marcellus Shale is a major geologic formation underlying significant portions of New York, Ohio, Pennsylvania, and West Virginia. Although it is a very tight formation, it contains a massive quantity of natural gas, thus making it of great economic importance. This paper covers the geology of the Marcellus Shale (extent, depth, gas producing potential, properties, etc.), the techniques used to produce the gas, and the potential for Carbon Capture and Storage (CCS) in the Marcellus Shale or adjacent formations. Because of the low permeability of shale units, hydraulic fracturing and horizontal drilling were developed in the Barnett Shale of Texas during the 1990's; these were the key enabling technologies that made recovery of shale gas economically viable. These technologies have been applied to the Marcellus Shale and other shale gas basins. In addition to gas production from the Marcellus Shale and other gas shale basins in the U.S., this paper discusses the impact of shale gas exploration and production on the potential for CCS in the Marcellus and other units in the Appalachian Basin.

Chapter 5- The presence of natural gas—primarily methane—in the shale layers of sedimentary rock formations that were deposited in ancient seas has been recognized for many years. The difficulty in extracting the gas from these rocks has meant that oil and gas companies have historically chosen to tap the more permeable sandstone or limestone layers which give up their gas more easily.

In: Marcellus Shale and Shale Gas
Editor: Gabriel L. Navarro ISBN: 978-1-61470-173-6

Chapter 1

WATER RESOURCES AND NATURAL GAS PRODUCTION FROM THE MARCELLUS SHALE*

Daniel J. Soeder[1] and William M. Kappel[2]

INTRODUCTION

The Marcellus Shale is a sedimentary rock formation deposited over 350 million years ago in a shallow inland sea located in the eastern United States where the present-day Appalachian Mountains now stand (de Witt and others, 1993). This shale contains significant quantities of natural gas. New developments in drilling technology, along with higher wellhead prices, have made the Marcellus Shale an important natural gas resource.

The Marcellus Shale extends from southern New York across Pennsylvania, and into western Maryland, West Virginia, and eastern Ohio (fig. 1). The production of commercial quantities of gas from this shale

---

* This is an edited, reformatted and augmented version of the United States Department of the Interior, United States Geological Survey publication, Fact Sheet 2009–3032, dated May 2009.

[1] U.S. Geological Survey, MD-DE-DC Water Science Center, 5522 Research Park Drive, Baltimore, MD 21228.

[2] U.S. Geological Survey, New York Water Science Center, 30 Brown Road, Ithaca, NY 14850.

requires large volumes of water to drill and hydraulically fracture the rock. This water must be recovered from the well and disposed of before the gas can flow. Concerns about the availability of water supplies needed for gas production, and questions about wastewater disposal have been raised by water-resource agencies and citizens throughout the Marcellus Shale gas development region. This Fact Sheet explains the basics of Marcellus Shale gas production, with the intent of helping the reader better understand the framework of the water-resource questions and concerns.

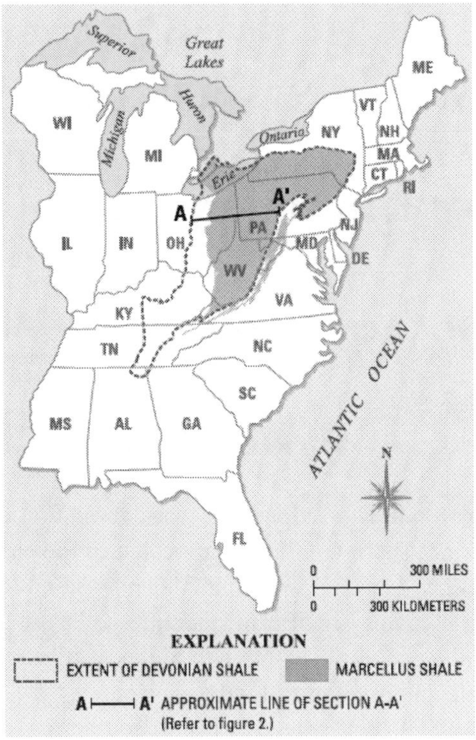

Figure 1. Distribution of the Marcellus Shale (modified from Milici and Swezey, 2006).

What is the Marcellus Shale?

The Marcellus Shale forms the bottom or basal part of a thick sequence of Devonian age, sedimentary rocks in the Appalachian Basin. This sediment was deposited by an ancient river delta, the remains of which now form the Catskill

Mountains in New York (Schwietering, 1979). The basin floor subsided under the weight of the sediment, resulting in a wedge-shaped deposit (fig. 2) that is thicker in the east and thins to the west. The eastern, thicker part of the sediment wedge is composed of sandstone, siltstone, and shale (Potter and others, 1980), whereas the thinner sediments to the west consist of finer-grained, organic-rich black shale, interbedded with organic-lean gray shale. The Marcellus Shale was deposited as an organic-rich mud across the Appalachian Basin before the influx of the majority of the younger Devonian sediments, and was buried beneath them.

WHY IS THE MARCELLUS SHALE AN IMPORTANT GAS RESOURCE?

Organic matter deposited with the Marcellus Shale was compressed and heated deep within the Earth over geologic time, forming hydrocarbons, including natural gas. The gas occurs in fractures, in the pore spaces between individual mineral grains, and is chemically adsorbed onto organic matter within the shale (Soeder, 1988). To produce commercial amounts of natural gas from such fine-grained rock, higher permeability flowpaths must be intercepted or created in the formation. This is generally done using a technique called hydraulic fracturing or a *"hydrofrac,"* where water under high pressure forms fractures in the rock, which are propped open by sand or other materials to provide pathways for gas to move to the well. Petroleum engineers refer to this fracturing process as *"stimulation."*

From the mid-1970s to early 1980s, the U.S. Department of Energy (DOE) funded the Eastern Gas Shales Project (EGSP) to develop new technology in partnership with industry that would advance the commercial development of Devonian shale gas (Schrider and Wise, 1980). Goals of the project included assessing the size of the resource, estimating recoverable gas, and determining the most effective technology for gas extraction. The EGSP shale stimulation experiments tested a wide variety of hydrofracs and other techniques. Results were somewhat uneven, and DOE concluded that stimulation alone was generally insufficient to achieve commercial shale gas production (Horton, 1982). It was suggested that better success could be obtained by targeting specific formations in specific locations. The EGSP results did indicate that if the hydraulic fractures were able to intercept sets of existing, natural fractures within the shale (fig. 3), a network of flowpaths could be created.

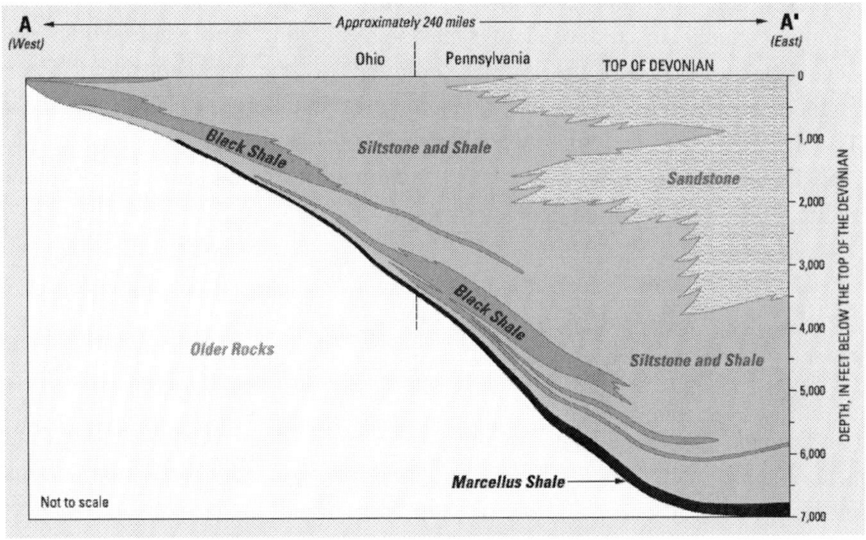

Figure 2. West to east line of section A-A' of Middle and Upper Devonian rocks in the Appalachian Basin. The Marcellus Shale is the lowest unit in the sequence (modified from Potter and others, 1980).

Figure 3. Marcellus Shale drill core from West Virginia, 3.5 inches in diameter, containing a calcite-filled vertical natural fracture. Photograph by Daniel Soeder, USGS.

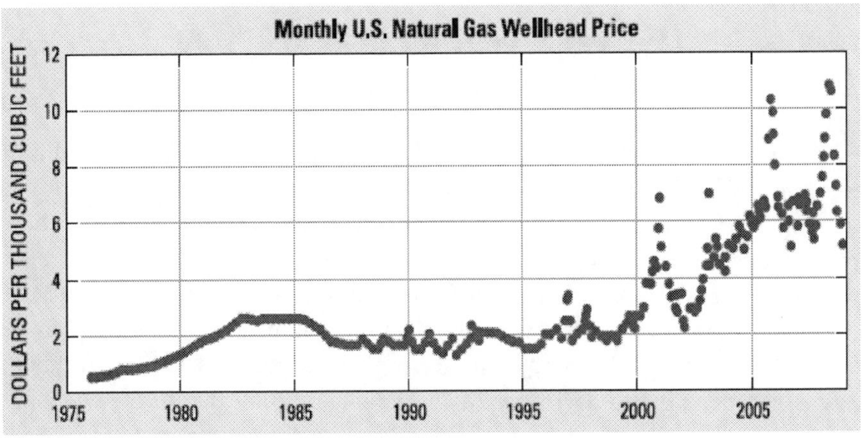

Figure 4. Wellhead price of natural gas since the mid-1970s to January 2009. [Source U.S. Energy Information Administration, 2009].

In the mid-1980s, the Institute of Gas Technology (IGT) in Chicago performed some laboratory analyses on EGSP shale samples (Soeder, 1988). Published IGT lab measurements found that a *"gas-in-place"* value for the Marcellus Shale at pressures representative of production depths may be as high as 26.5 standard cubic feet of gas per cubic foot of rock (Soeder, 1988). This greatly exceeded earlier gas-in-place estimates for Devonian shale by the National Petroleum Council (1980), which ranged from 0.1 to 0.6 standard cubic feet of gas per cubic foot of rock. Although IGT analyzed only one sample of Marcellus Shale, the large volumes of gas now being produced from this formation substantiate the early discovery of significant gas reserves.

In 2008, two professors at Pennsylvania State University and the State University of New York (SUNY) Fredonia estimated that about 50 TCF (trillion cubic feet) of recoverable natural gas could be extracted from the Marcellus Shale (Engelder and Lash, 2008). In November 2008, on the basis of production information from Chesapeake Energy Corporation, the estimate of recoverable gas from the Marcellus Shale was raised to more than 363 TCF (Esch, 2008). The United States uses about 23 TCF of natural gas per year (U.S. Energy Information Administration, 2009), so the Marcellus gas resource may be large enough to supply the needs of the entire Nation for roughly 15 years at the current rates of consumption.

Why is the Marcellus Shale Being Developed Now?

Low prices for natural gas, and ineffective production technology did little to spark interest in Devonian shale gas in the 1990s, despite the publication of the IGT Marcellus gas-in-place estimates. Two factors working together have promoted the current high levels of interest in the Marcellus Shale. First, wellhead prices for gas have risen from values of less than $2.00 per MCF (thousand cubic feet) in the 1980s (fig. 4) to a peak of $10.82 per MCF in the summer of 2008 (U.S. Energy Information Administration, 2009). Although prices had declined to $5.15 per MCF in January 2009 due to the economic downturn, they are still substantially higher than a decade earlier.

The second factor that spurred interest in the development of the Marcellus Shale is a new application of an existing drilling technology known as *directional drilling*, which involves steering a downhole drill bit in a direction other than vertical. An initially vertical drillhole is slowly turned 90 degrees to penetrate long horizontal distances, sometimes over a mile, through the Marcellus Shale bedrock. Hydraulic fractures are then created into the rock at intervals from the horizontal section of the borehole, allowing a substantial number of high-permeability pathways to contact a large volume of rock (fig. 5).

According to Range Resources (2008), one of the first major horizontal drillers of Marcellus Shale, these wells typically produce gas at a rate of about 4 MMCF (million cubic feet) per day. Over its lifetime, each horizontal well on an 80-acre surface spacing can be expected to produce a total of about 2.5 BCF (billion cubic feet) of gas at an estimated production cost of $1.00 per MCF (Chernoff, 2008).

What are the Water-Resource Concerns About Developing Natural Gas Wells in the Marcellus Shale?

Substantial amounts of water are required for the drilling and stimulation of a Marcellus Shale gas well. Fluids recovered from the well, including the liquids used for the hydrofrac, and any produced formation brines, must be treated and disposed of properly. Three important water-resource concerns related to Marcellus Shale gas production are:

- supplying water for well construction without impacting local water resources,
- avoiding degradation of small watersheds and streams as substantial amounts of heavy equipment and supplies are moved around on rural roads, and
- determining the proper methods for the safe disposal of the large quantities of potentially contaminated fluids recovered from the wells.

These concerns are discussed in more detail in the following sections.

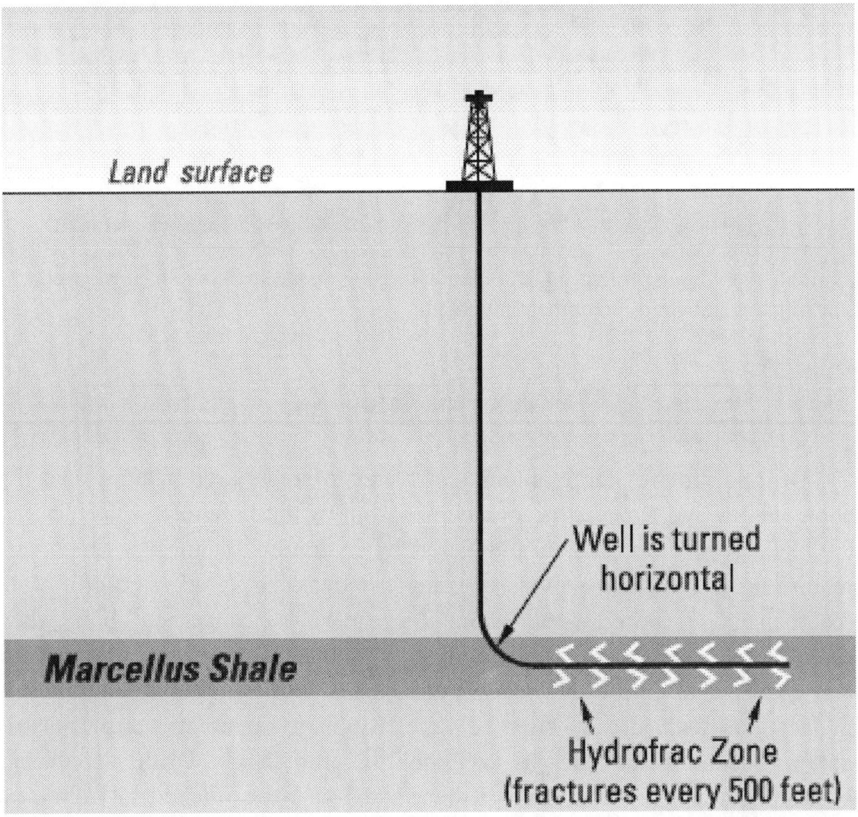

Figure 5. Combination of directional drilling and hydraulic fracturing technology used for gas production from the Marcellus Shale in the Appalachian Basin (modified from http://geology.com/articles/marcellus-shale.shtml).

Figure 6. A hydraulic fracturing stimulation in 2007 on a Marcellus Shale gas well showing the amount of equipment involved.

WATER SUPPLY

Drilling requires large amounts of water to create a circulating mud that cools the bit and carries the rock cuttings out of the borehole. After drilling, the shale formation is then stimulated by hydraulic fracturing, which may require up to 3 million gallons of water per treatment (Harper, 2008). Many regional and local water management agencies are concerned about where such large volumes of water will be obtained, and what the possible consequences might be for local water supplies. Under drought conditions, or in locations with already stressed water supplies, obtaining the millions of gallons needed for a shale gas well could be problematic. Drillers could face substantial transportation costs if the water has to be trucked in from great distances.

Similar shale gas operations in the Barnett Shale of Texas have obtained hydrofrac water largely from groundwater sources (Byrd, 2007). Water-supply concerns over the Barnett Shale drilling have been brought up in the past (see, for example, Francis, 2007). Texas State and County agencies now closely monitor volumes of water used during drilling, and a consortium of

Barnett Shale drilling companies have developed best management practices for water conservation, with the goal of keeping the pace of drilling and production activities within the bounds of sustainable water use. Similar steps have been discussed in Marcellus Shale gas production areas, but not yet fully implemented.

TRANSPORTING FLUIDS AND SUPPLIES

Large hydrofrac treatments often involve moving large amounts of equipment, vehicles, and supplies into remote areas (fig. 6). Transporting all of this to drill sites over rural Appalachian Mountain roads could potentially cause erosion, and threaten local small watersheds with sediment. Drill pad and pipeline construction also have the potential to cause similar problems. Of equal concern is the possibility for spills or leaks into water bodies as the fluids and chemical additives are transported and handled. Little is known about how a Marcellus Shale drilling "boom" might adversely affect the land, streams, and available water supplies in the Appalachian Basin. Even under current Marcellus gas production levels, complaints of rural road damage and traffic disruption from drilling equipment have been received, indicating that this could be a significant problem if carried out across thousands of active drill sites.

WASTEWATER DISPOSAL

For gas to flow out of the shale, nearly all of the water injected into the well during the hydrofrac treatment must be recovered and disposed of. In addition to the problem of dealing with large bulk volumes of liquid waste, contaminants in the water may complicate wastewater treatment. Whereas the percentage of chemical additives in a typical hydrofrac fluid is commonly less than 0.5 percent by volume, the quantity of fluid used in these hydrofracs is so large that the additives in a three million gallon hydrofrac job, for example, would result in about 15,000 gallons of chemicals in the waste.

Hydrofrac fluids are often treated with proprietary chemicals to increase the viscosity to a gel-like consistency that enables the transport of a *proppant*, usually sand, into the fracture to keep it open after the pressure is released (fig. 7). The viscosity of these fluids then breaks down quickly after completion of the hydrofrac, so they can be easily removed from the ground. The chemical formulations required to achieve this are highly researched and closely guarded, and finding out exactly what is in these fluids may present a challenge. The data publicly available on Marcellus Shale hydrofrac treatments indicate that a *slickwater frac* works best on this formation (Harper, 2008). These types of hydrofracs employ linear gels and friction reducers in the water, and utilize only small amounts of proppant, relying instead on fracture surface roughness to hold it open (Rushing and Sullivan, 2007). The potential problems for local wastewater treatment facilities caused by proprietary chemical additives in hydrofrac fluid are unclear.

Along with the introduced chemicals, hydrofrac water is in close contact with the rock during the course of the stimulation treatment, and when recovered may contain a variety of formation materials, including brines, heavy metals, radionuclides, and organics that can make wastewater treatment difficult and expensive. The formation brines often contain relatively high concentrations of sodium, chloride, bromide, and other inorganic constituents, such as arsenic, barium, other heavy metals, and radionuclides that significantly exceed drinking-water standards (Harper, 2008).

The current disposal practice for Marcellus Shale liquids in Pennsylvania requires processing them through wastewater treatment plants, but the effectiveness of standard wastewater treatments on these fluids is not well understood. In particular, salts and other dissolved solids in brines are not usually removed successfully by wastewater treatment, and reports of high salinity in some Appalachian rivers have been linked to the disposal of Marcellus Shale brines (Water and Wastes Digest, 2008). Another disposal option involves re-injecting the hydrofrac fluids back into the ground at a shallower depth. This is a common practice in the Barnett Shale production area of Texas, and has been utilized for some Marcellus wells drilled in West Virginia (Kasey, 2008). Concerns in Appalachian States about the possible contamination of drinking-water supply aquifers has limited the practice of re-injecting Marcellus fluids, however. Another option might be to inject the waste fluid into deeper formations below the Marcellus Shale that are not used as aquifers, such as the Oriskany or Potsdam Sandstones. A third disposal process used in Texas places the wastewater into an open tank to evaporate. The solids that remain behind are then disposed of as dry waste. Although this may be an effective technique in the deserts of the American Southwest, its usefulness in the humid climate of the Appalachians is questionable. A systematic study of the options for Marcellus Shale waste fluid treatment, disposal, or recycling could help to determine the best available procedures.

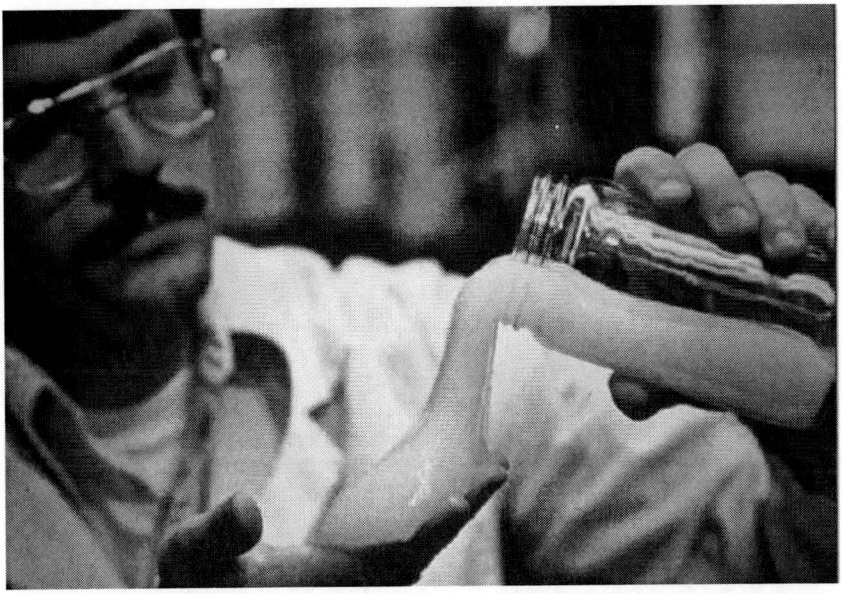

Figure 7. Example of a gel used in hydrofracturing to carry proppant into a fracture. Photograph by Daniel Soeder, USGS.

SUMMARY

Natural gas is an abundant, domestic energy resource that burns cleanly, and emits the lowest amount of carbon dioxide per calorie of any fossil fuel. The Marcellus Shale and other natural gas resources in the United States are important components of a national energy program that seeks both greater energy independence and greener sources of energy. Marcellus gas development has begun in the northern Appalachian Basin, with significant lease holdings throughout Pennsylvania, West Virginia, southern New York, western Maryland, and eastern Ohio. Because of questions related to water supply and wastewater disposal, however, many state agencies have been cautious about granting permits, and some states have placed moratoriums on drilling until these issues are resolved. At the same time, gas companies, drillers, and landowners are eager to move forward and develop the resource.

While the technology of drilling directional boreholes, and the use of sophisticated hydraulic fracturing processes to extract gas resources from tight rock have improved over the past few decades, the knowledge of how this extraction might affect water resources has not kept pace. Agencies that manage and protect water resources could benefit from a better understanding of the impacts that drilling and stimulating Marcellus Shale wells might have on water supplies, and a clearer idea of the options for wastewater disposal.

References

Byrd, C.L., 2007, Updated evaluation for the Central Texas Trinity Aquifer Priority Groundwater Management Area, Priority Groundwater Management Area File Report: Austin, TX, Texas Commission on Environmental Quality, Water Rights Permitting and Availability Section, Water Supply Division, 160 p.

Chernoff, Harry, 2008, Investing in the Marcellus Shale, article on Seeking Alpha (investing) website, accessed March 17, 2009 at *http:// seekingalpha.com/article/68716- investing-in-the-marcellus-shale.*

de Witt, Wallace, Jr., Roen, J.B., and Wallace, L.G., 1993, Stratigraphy of Devonian black shales and associated rocks in the Appalachian basin, *in* Roen, J.B., and Kepferle, R.C., eds., Petroleum geology of the Devonian and Mississippian black shale of eastern North America: U.S. Geological Survey Bulletin 1909–B, p. B1–B57.

Engelder, Terry, and Lash, Gary, 2008, Unconventional natural gas reservoir could boost U.S. supply, Penn State Live, accessed January 10, 2009 at *http://live.psu.edu/ story/28116.*

Esch, Mary, 2008, Estimated gas yield from Marcellus shale goes up: Albany, NY, Associated Press, November 4, 2008, accessed March 17, 2009 at *http://www.ibtimes.com/articles/20081104/estimated-gasyield-from-marcellus-shale-goes-up. htm.*

Francis, R., 2007, Data on water use in the Barnett Shale begins to surface: Fort Worth, TX, Fort Worth Business Press, March 12, 2007, accessed March 16, 2009 at *http://www.fwbusinesspress.com/display.php?id=5838.*

Harper, J.A., 2008, The Marcellus Shale — an old "new" gas reservoir, *in* Pennsylvania Geology: Pennsylvania Department of Conservation and Natural Resources, v. 38, no. 1, 20 p.

Horton, A.I., 1982, 95 stimulations in 63 wells — DOE reports on comparative analysis of stimulation strategy in eastern gas shales: Columbus, OH, Northeast Oil Reporter, p. 63–73, February 1982.

Kasey, Pam, 2008, New drilling efforts raise questions: Charleston, WV, The State Journal, August 14, 2008, accessed March 16, 2009 at *http:// statejournal.com/story.cfm?func=vie wstory&storyid=42542.*

Milici, R.C., and Swezey, C.S., 2006, Assessment of Appalachian Basin oil and gas resources: Devonian shale — Middle and Upper Paleozoic total petroleum system: U.S. Geological Survey Open-File Report 2006–1237, 70 p., accessed April 22, 2009 at *http://pubs.usgs.gov/of/2006/1237/*.

National Petroleum Council, 1980, Unconventional gas sources, Volume III: Devonian shale: Report prepared for the U.S. Secretary of Energy, June 1980, National Petroleum Council, Washington, D.C. 20006, 252 p., accessed March 16, 2009 at *http:// www.npc.org/*.

Potter, P.E., Maynard, B., and Pryor, W.A., 1980, Final report of special geological, geochemical, and petrological studies of the Devonian shales of the Appalachian Basin, prepared for U.S. Department of Energy under contract DE–AC21–76MC05201: Cincinnati, OH, University of Cincinnati, January 1980, [variously paged].

Range Resources, 2008, Range provides Marcellus Shale update, corporate press release: Fort Worth, TX, Range Resources Corporation, July 14, 2008, accessed March 16, 2009 at *http://www.rangeresources. com/PressReleases.asp*.

Rushing, J.A., and Sullivan, R.B., 2007, Improved water frac increases production: Houston, TX, E&P Magazine, Hart Energy Publishing, September 2007, accessed January 28, 2008 at *http://www.epmag.com/ archives/features/661.htm*.

Schrider, L.A., and Wise, R.L., 1980, Potential new sources of natural gas: Journal of Petroleum Technology, v. 32, no. 4, p. 703–716, DOI 10.2118/7628–PA.

Schwietering, J.F., 1979, Devonian shales of Ohio and their eastern and southern equivalents: Morgantown, WV, Report METC/CR–79/2, prepared for U.S. Department of Energy under contract EY–76–C–05–5199, West Virginia Geologic and Economic Survey, January 1979.

Soeder, D.J., 1988, Porosity and permeability of eastern Devonian gas shale: Society of Petroleum Engineers Formation Evaluation, Society of Petroleum Engineers, v. 3, no. 2, p. 116–124, DOI 10.2118/15213–PA.

U.S. Energy Information Administration, 2009, U.S. natural gas total consumption, accessed April 15, 2009 at *http://www.eia.doe.gov/ oil_gas/natural_gas/info_glance/ natural_gas.html*.

Water and Wastes Digest, 2008, Pennsylvania DEP investigates elevated TDS in Monongahela River, accessed April 10, 2009 at *http:// www.wwdmag.com/Pennsylvania-DEP-Investigates-Elevated-TDS-in-Monongahela-River-NewsPiece16950*.

In: Marcellus Shale and Shale Gas
Editor: Gabriel L. Navarro

ISBN: 978-1-61470-173-6

Chapter 2

WATER MANAGEMENT TECHNOLOGIES USED BY MARCELLUS SHALE GAS PRODUCERS*

United States Department of Energy

Submitted by:
John A. Veil
Argonne National Laboratory
Argonne, IL

Prepared for:
United States Department of Energy
National Energy Technology Laboratory

About Argonne National Laboratory

Argonne is a U.S. Department of Energy laboratory managed by UChicago Argonne, LLC under contract DE-ACO2-06CH11357. The Laboratory's main facility is outside Chicago, at 9700 South Cass Avenue, Argonne, Illinois 60439. For information about Argonne and its pioneering science and technology programs, see www.anl.gov.

---

* This is an edited, reformatted and augmented version of a Office of Fossil Energy publication, DOE Award No.: FWP 49462, dated July 2010.

DISCLAIMER

prepared for
U.S. Department of Energy, Office of Fossil Energy,
National Energy Technology Laboratory

prepared by
J.A. Veil
Environmental Science Division,
Argonne National Laboratory
July 2010

INTRODUCTION

Natural gas represents an important energy source for the United States. According to the U.S. Department of Energy's (DOE's) Energy Information Administration (EIA), about 22% of the country's energy needs are provided by natural gas. Historically, natural gas was produced from conventional vertical wells drilled into porous hydrocarbon-containing formations. During the past decade, operators have increasingly looked to other unconventional sources of natural gas, such as coal bed methane, tight gas sands, and gas shales.

Figure 1 shows EIA projections of the source of natural gas supplies through 2030. Unconventional gas supplies are anticipated to play an increasingly important role. Some of the busiest and most productive oil and gas activities in the country today are shale gas plays.

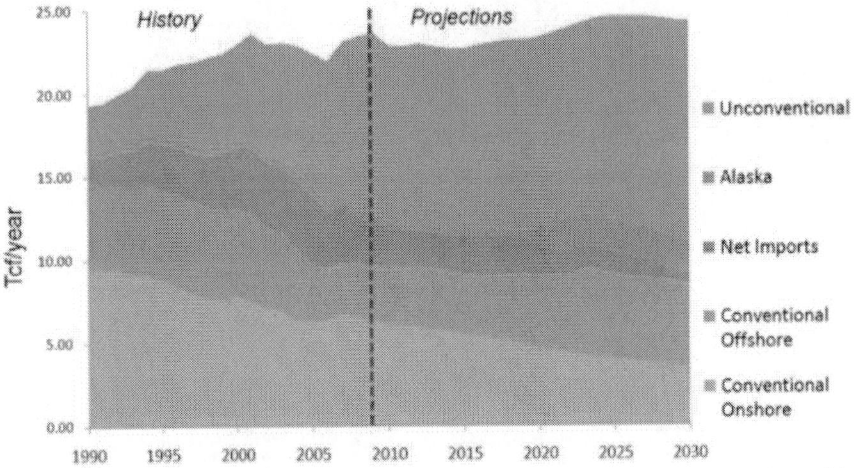

Source: DOE/EIA Annual Energy Outlook 2009. Note that Tcf refers to trillion cubic feet.

Figure 1. U.S. Natural Gas Supply by Source.

Shale Gas Resources in the United States

Important shale gas formations are found in many parts of the United States, as shown on the map in Figure 2. Much of the early rapid growth in shale gas production took place in the Barnett Shale formation near Fort Worth, Texas. As the technology evolved, operators began to explore other large shale formations in other parts of the country. The most active shales to date are the Barnett Shale, the Fayetteville Shale, the Antrim Shale, the Haynesville Shale, the Marcellus Shale, and the Woodford Shale. A 2009 Shale Gas Primer, sponsored by DOE, includes a chart showing the gas production from several major shale gas formations (GWPC and ALL 2009 – see page 10).

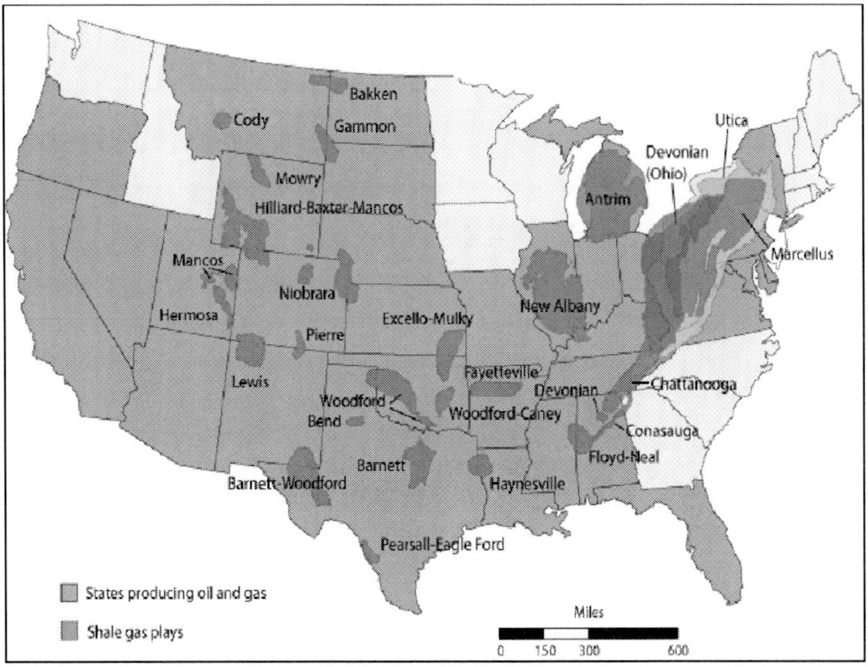

Source: Provided by staff from DOE's Office of Fossil Energy.

Figure 2. U.S. Shale Gas Plays.

Technologies that Enable Shale Gas Production

Unlike conventional natural gas, which has been produced for more than a century, shale gas is more difficult to remove from the ground. Shale formations contain very tight rock, with less pore space than traditional oil and gas formations, such as sandstone and limestone. Conventional drilling and production methods typically cannot produce enough natural gas from shale formations to make the wells economically viable.

Because of the tight nature of the shale formations, gas producers have relied on more advanced technologies in order to extract sufficient volumes of natural gas to make the wells profitable. The two technologies that have lead the way in allowing economical production of shale gas are horizontal drilling and multistage hydraulic fracturing.

Horizontal drilling is an advanced technology that allows a well to be drilled vertically to a desired depth, then turned sideways to reach out hundreds to thousands of feet laterally. Horizontal drilling is critical for producing shale gas because it creates a well that penetrates through a long section of the shale formation, allowing for the collection of gas throughout a much longer horizontal run.

The hydraulic fracturing process (a “frac job”) injects water, sand, and other ingredients at very high pressure into the well. The high pressure creates small fractures in the rock that extend out as far as 1,000 feet away from the well. After the fractures are created, the pressure is reduced. Water from the well returns to the surface (known as flowback), but the sand grains remain in the rock fractures, effectively propping the fractures open and allowing the gas to move. Frac jobs on traditional vertical wells are usually done in one stage. However, given the length of horizontal wells, the frac jobs are often conducted in limited linear sections of the well known as stages. In a long horizontal well, stages are fractured sequentially, beginning with the outermost section of the well. There is a useful video animation clip at http://www.pamarcellus.com/web that shows how horizontal wells are drilled and then hydraulically fractured.

Various types of “frac fluids” and additives have been used. Most frac fluid used in shale gas wells consists of water, a proppant (generally sand), a friction reducing agent (“slick water” — water containing some surfactant additives to help the flow-back water return from the well at the end of the frac job), and other chemicals used to protect the well and to optimize performance. GWPC and ALL (2009) lists the major additives, the types of chemicals found in the additives, and their functions.

DOE/NETL Research Program

DOE’s National Energy Technology Laboratory (NETL) administers an Environmental Program that aims to find solutions to environmental concerns by focusing on the following program elements:

1) Produced water and fracture flowback water management, particularly in gas shale development areas,
2) Water resource management in oil and gas basins,
3) Air quality issues associated with oil and gas exploration and production (E&P) activities,
4) Surface impact issues associated with E&P activities,
5) Water resource management in Arctic oil and gas development areas,

6) Decision making tools that help operators balance resource development and environmental protection, and
7) Online information and data exchange systems that support regulatory streamlining.

There are currently 27 extramural projects in the Environmental Program, with a total value of roughly $32 million (not including participant cost-share). Approximately $10 million of this total is directed toward projects led by industry, $9 million to projects led by universities, $11 million to state agencies and national non-profit organizations, and $2 million to national laboratories for technical support to other project partners. The project portfolio is balanced between projects focused on technology development, data gathering, and development of data management software and decision support tools.

Some of these projects are referenced in this report. Program and individual project information can be found at the following NETL links:

- Technology Solutions for Mitigating Environmental Impacts of Oil and Gas E&P Activity
  http://www.netl.doe.gov/publications/factsheets/program/Prog101.pdf
- Natural Gas and Petroleum Projects, Environmental Solutions, Produced Water Management
  http://www.netl.doe.gov/technologies/oil

WATER ISSUES ASSOCIATED WITH SHALE GAS PRODUCTION

Water plays a role in different aspects of shale gas production. Three important water issues are discussed in this chapter.

Stormwater Runoff from Disturbed Areas

In order to create an area for drilling a new well, the operator clears and grades an area that can accommodate one or more wellheads; several pits for holding water, drill cuttings, and used drilling fluids; and space for the many trucks used to complete a frac job. Typically, this space will be 3 to 5 acres in size, plus any area disturbed to create an access road from the nearest public road to the well pad. Most of the figures in this chapter are photos taken by the author at several different Marcellus Shale well sites in southwestern Pennsylvania on a rainy day in May 2009 (photos from other locations are identified).

Figure 3 shows a well pad while the well is being drilled, Figure 4 shows a pad while the well is undergoing a frac job, and Figure 5 shows a pad with a completed wellhead. Figures 6 and 7 show examples of access roads at well sites in the same area. Figure 6 is taken at a recently completed well, while Figure 7 is taken at a much older well.

Figure 3. Well Pad Showing Drilling Rig.

Figure 4. Well Pad Showing Equipment Used for Frac Job.

Figure 5. Well Pad Showing Completed Wellhead.

Figure 6. Access Road at Recently Completed Well.

Figure 7. Access Road at Older Well.

These photos give an idea of the amount of disturbed land there is at a well site. Most operators employ appropriate management practices to control stormwater runoff. Figures 8 through 10 show some of the stormwater management structures that are used to capture offsite stormwater and divert it around the disturbed well pad area. This reduces the amount of water that carries sediment. The water falling on disturbed areas of the site can be controlled through the application of gravel to the well pad and road surfaces or through onsite collection pits. The Pennsylvania Department of Environmental Protection (PADEP), Bureau of Oil and Gas Management website contains presentations from a January 2010 training course (http://www.dep.state sTraining.htm). Several of the presentations relate to erosion and sediment control plans. The information and graphics contained in the presentation are useful.

Figure 8. Stormwater Diversion Ditch to Collect Offsite Water.

Figure 9. Lower End of Stormwater Diversion Ditch.

Figure 10. Stormwater Control Structure.

Water Supply for Drilling and to Make up Frac Fluids

The second important water issue involves finding an adequate and dependable supply of water to support well drilling and completion activities. Water used for drilling and making up frac fluids can come from several sources: surface water bodies, groundwater, municipal potable water supplies, or reused water from some other water source (most commonly this is flowback water from a previously fractured well).

GWPC and ALL (2009) provide estimates of water requirements for four of the major shale gas plays. The water required for drilling a typical shale gas well ranges from 1,000,000 gallons in the Haynesville Shale to 60,000 gallons in the Fayetteville Shale, depending on the types of drilling fluids used and the depth and horizontal extent of the wells. The Marcellus Shale drilling volume falls near the lower end of this range at 80,000 gallons per well. The volume needed to fracture a well is considerably larger. According to GWPC and ALL (2009), the frac fluid volume ranges from 3,800,000 gallons per well in the Marcellus Shale to 2,300,000 gallons per well in the Barnett Shale.

Another source of information on the amount of water used per well is a presentation given by a representative of the Susquehanna River Basin Commission (SRBC) on volumes of water withdrawn for Marcellus Shale gas well development. A large portion of the Marcellus Shale underlies the Susquehanna River basin watershed. Any water usage within the watershed is subject to oversight by the SRBC. Hoffman (2010) notes, that as of January 2010, the SRBC had data for 131 wells. The total volume of water withdrawn through that date is 262 million gallons, with 45% coming from public water supplies and the other 55% coming from surface water sources. The average total volume of fluid used per well is 2.7 million gallons, with 2.2 million gallons of that coming from freshwater sources and 0.5 million gallons coming from recycled flowback water. No information was provided by Hoffman (2010) concerning whether the wells in the SRBC data set were vertical or horizontal wells (a vertical well requires much less water for a frac job than does a horizontal well).

Water can be brought to the site by numerous tank trucks or, where another source of water is available within a mile or so, it can be piped to the site. Figure 11 shows tank trucks similar to those used to haul water. The photo on the left of a large semi-style tank truck was taken in the Barnett Shale region of Texas, and the photo on the right of two smaller tank trucks was taken in western Pennsylvania. Figures 12 through 14 show pipes conveying water to a well undergoing a frac job. Figures 12 and 13 were taken in southwestern Pennsylvania. Figure 12 shows a pump withdrawing water from a storage pond and the rubber hose used to convey it up a hill to a well site. Figure 13 shows the same pipe near the top of the hill approaching the well site.

Figure 14 shows aluminum pipes used to convey water at a site in the Fayetteville Shale region of Arkansas.

Figure 11. Tank Trucks Similar to Those Used to Deliver Water to a Well Site.

Figure 12. Pumping Unit Used to Move Water to Next Well Site.

Figure 13. Pipe Shown at the top of a Hill after Pumping.

Figure 14. Aluminum Pipe Used to Convey Water.

Management of Water Flowing to the Surface from the Well

The third important water issue involves managing the water that comes to the surface from the gas well. During the frac job, the operator injects a large volume of water into the formation. Once the frac job is finished, the pressure is released, and a portion of the injected water flows back to the surface in the first few days to weeks. This water is referred to as flowback or flowback water. Over a much longer period of time, additional water that is naturally present in the formation (i.e., produced water) continues to flow from the well. While some authors consider flowback to be just one part of the produced water, this report distinguishes flowback from the ongoing produced water. Both flowback and produced water typically contain very high levels of total dissolved solids (TDS) and many other constituents. Over an extended period of time, the volume of produced water from a given well decreases.

Not all of the injected frac fluid returns to the surface. GWPC and ALL (2009) report that from 30% to 70% of the original frac fluid volume returns as flowback. However, anecdotal reports from Marcellus operators suggest that the actual percentage is at or below the lower end of that range. The rest of the water remains in pores within the formation. The SRBC data set described in the previous section shows that about 13.5% of the injected frac fluid is recovered (Hoffman 2010).

Operators must manage the flowback and produced water in a cost-effective manner that complies with state regulatory requirements. The primary options are:

- Inject underground through a disposal well (onsite or offsite),
- Discharge to a nearby surface water body,
- Haul to a municipal wastewater treatment plant (often referred to as a publicly owned treatment works or POTW),
- Haul to a commercial industrial wastewater treatment facility, and
- Reuse for a future frac job either with or without treatment.

Chapter 3 describes each of these different processes in more detail and identifies those options that are actually being used by gas operators in the Marcellus Shale region.

WATER MANAGEMENT TECHNOLOGIES USED IN THE MARCELLUS SHALE

The last portion of Chapter 2 describes the range of potential options for managing flowback and produced water from shale gas wells. This chapter reviews these options and identifies which of the options are currently being used by Marcellus Shale gas operators.

Data Collection Approach

In order to identify the water management options that are actually being used, Argonne National Laboratory (Argonne) contacted the Marcellus Shale Coalition, a group of companies involved with natural gas production in the Marcellus Shale region. Although the Coalition is primarily focused on gas development in Pennsylvania, many of the companies are currently working in or plan to work in the other Marcellus Shale states too. A representative of the Coalition provided a list of eight major companies operating in the Marcellus Shale along with the names of contact persons. Argonne wrote to each of these companies, plus another three companies with which Argonne is currently working on other Marcellus water projects. Argonne received replies from eight companies, including one reply that indicated the company's water management practices were confidential at that point.

In addition, the author made visits during May 2010 to four commercial industrial wastewater treatment facilities that accept water from Marcellus Shale gas wells. These facilities employ different processes to treat the flowback and produced water. The facilities are described in Chapter 4.

Underground Injection

GWPC and ALL (2009) list the water management options employed at several different shale gas plays. All of the plays employ injection wells as a *primary* means of disposal except for the Marcellus. Few, if any, onsite injection wells are used in Pennsylvania or New York, nor are there any commercial disposal wells used for Marcellus Shale flowback and produced water located in these states.

Where injection is available (e.g., at the other shale gas plays and in portions of Ohio or West Virginia for Marcellus Shale flowback and produced water), the injection wells can be either onsite wells operated by the gas producer or offsite third-party commercial disposal wells. To give readers a sense of what a commercial disposal well looks like, Figure 15 shows a commercial flowback and produced water disposal well in the Barnett Shale region of Texas. Flowback and produced water are delivered by tank truck and are transferred into the storage tanks. As necessary, the flowback and produced water are injected into a deep formation that has sufficient porosity and injectivity to accept the water.

Figure 15. Injection Well and Tank Battery at Commercial Disposal Facility in Texas.

At least some of the Marcellus Shale gas operators are sending flowback and produced water to commercial disposal wells located in Ohio. The author contacted the Ohio Department of Natural Resources, Division of Mineral Resources Management (DMRM), to learn which commercial wastewater disposal companies were operating injection wells in eastern and central Ohio to receive flowback and produced water from Pennsylvania. Tom Tomastik, an Underground Injection Control program manager with the Ohio DMRM, replied:

> "Ohio does not distinguish between commercial and non-commercial Class II injection wells. So basically, it's the operator who determines if they will take other operator's oil and gas fluids. Although we track brine hauling from cradle to grave, it would be a huge process right now to go through and delineate the fluids from the Marcellus. Starting on June 30th, however, passage of Ohio Senate Bill 165 will require a 20 cent per barrel fee for out-of-district oilfield fluids to be paid to us, so we will start tracking out-of-state fluids more closely (mostly Marcellus) then." (Tomastik 2010).

Mr. Tomastik provided two tables listing Ohio's permitted injection wells. The first table listed companies that operate commercial disposal wells. The second table listed all permitted Class II injection wells (those used for injecting oil and gas fluids). This information was combined into Table 1, which shows the companies that operate commercial saltwater (flowback and produced water) injection wells and the counties in which those wells are located.

Additional information on 2009 injection volumes was provided by Gregg Miller of the Ohio DMRM (Miller 2010). That information, for each injection well, is shown in the right-hand column of Table 1.

Table 1. Commercial Saltwater Disposal Wells in Ohio and 2009 Injection Volume

County	Operator	Lease	2009 Disposal Volume (bbl/yr)
Ashtabula	B & B Oilfield Services, Inc.	Miller & Co. #3	75,212
Ashtabula	B & B Oilfield Services, Inc.	Clinton Oil SWIW #2	75,211
Athens	Carper Well Service	H. Ginsburg #1	49,163
Guernsey	David R. Hill, Inc.	Devco Unit #1	312,753
Guernsey	Dover Atwood Corp.	Kopolka #1	76,311
Guernsey	Arvilla Oilfield Services LLC	Slifko #1	14,835
Holmes	OOGC Disposal Company	Killbuck Disposal Well #1	164,202
Holmes	Mac Oilfield Service, Inc.	F. Hawkins #1	95,061
Licking	OOGC Disposal Company	Ronald F. Moran #1	190,437
Mahoning	Brineaway, Inc.	Salty Dog #1	26,940

County	Operator	Lease	2009 Disposal Volume (bbl/yr)
Mahoning	Brineaway, Inc.	Salty Dog #3	45,019
Mahoning	Brineaway, Inc.	Jenkins #1	52,181
Morgan	Broad Street Energy	Cook #2-A	15,442
Morrow	Fishburn Producing, Inc.	J.F. Mosher #1	11,585
Morrow	Fishburn Producing, Inc.	Fishburn #1	90,739
Morrow	Fishburn Producing, Inc.	Clinger Unit #1	47,250
Morrow	Fishburn Producing, Inc.	Power (Fegley) #1	20,853
Noble	Arvilla Oilfield Services LLC	H. Dudley #1	153,307
Noble	Triad Resources, Inc.	Warren Drilling #1	230,731
Noble	Carper Well Service	Bryan-Smith Unit #1	188,055
Perry	R.C. Poling	Rushcreek Partners et at #1	181,470
Portage	Ray Pander Trucking, Inc.	J. & D. Blazdek #2	81,206
Portage	Ray Pander Trucking, Inc.	Plum Creek #1	59,362
Portage	William S. Miller, Inc.	Wilcox #1	245,519
Portage	Salty's Disposal Wells, LP	Myers #1 Unit	206,535
Portage	Salty's Disposal Wells, LP	Groselle #2	176,989
Portage	B & B Oilfield Services, Inc.	Long #1	103,438
Stark	Ray Pander Trucking, Inc.	Belden & Blake Corp. SWDW #5	96,127
Stark	Ray Pander Trucking, Inc.	Ed Lyons (Genet) #1	171,790
Stark	Brineaway, Inc.	The Salty Dog #2	187
Stark	Brineaway, Inc.	J & E Walker #2	30,645
Stark	Brineaway, Inc.	Kolm #1	30,645
Trumbull	Ray Pander Trucking, Inc.	Eva Root Wolf #1	8,323
Trumbull	Ray Pander Trucking, Inc.	Wolf #2	104,443
Trumbull	Ray Pander Trucking, Inc.	Pander #1	113,941
Washington	Carper Well Service	Davis-Huffman #2	1,744

Table 1. (Continued)

County	Operator	Lease	2009 Disposal Volume (bbl/yr)
Washington	Carper Well Service	Davis-Huffman #3	9,180
Washington	Virco, Inc.	Helen Hall #1-19	121,978
Washington	Broad Street Energy	H.L. Flower #1	147,728
Washington	OOGC Disposal Company	Long Run Disposal #1	611,725
Wayne	Mac Oilfield Service, Inc.	Weldon Mohr #2	29,651
		Total Injected Volume	**4,467,913**

Note that Ohio has active oil and gas production from many wells that are not part of the Marcellus Shale. Some of the injection wells in Table 1 may dispose of water from these other oil and gas wells rather than from Marcellus Shale wells.

Figure 16 is a map of Ohio with the county boundaries delineated. Each county that has a permitted commercial saltwater disposal well is indicated with a star. Nearly all of the counties in which the disposal wells are located are in the eastern half of the state.

Argonne asked for similar information concerning commercial injection wells from the West Virginia Department of Environmental Protection (WVDEP) but did not receive any information. Puder and Veil (2006) report that three facilities in West Virginia operated commercial injection wells that accepted produced water in the time period between October 2005 and April 2006. However, only two of them provided information that could be used in that study. These were Base Petroleum in Charleston, West Virginia, and Danny Web Construction in Brenton, West Virginia.

Current Practices from Surveyed Gas Companies: Four of the surveyed gas companies indicated that they have sent produced water to commercial disposal wells. Three of the replies mentioned disposal wells in Ohio, while the fourth reply did not specify any location.

Source: Ohio Department of Transportation website at http://www.dot.state accessed May 28, 2010.

Figure 16. Ohio Counties in Which Commercial Disposal Wells Are Located (indicated by star).

Discharge to Surface Water Body

Many types of industrial wastewater are discharged to streams, rivers, and other surface water bodies. Permission to discharge wastewater is made through National Pollutant Discharge Elimination System (NPDES) permits issued by state agencies. However, discharging flowback or produced water directly from a well site presents various challenges. First, the water typically contains high levels of TDS (salinity) and other constituents that would require treatment.

In response to concern over flowback and produced water discharges, the PADEP, in April 2009, proposed a new strategy that would add effluent standards for oil and gas wastewaters of 500 mg/L for TDS, 250 mg/L for sulfates, 250 mg/L for chlorides, and 10 mg/L for total barium and total strontium.[1] On May 17, 2010, the Pennsylvania Environmental Quality Board approved the new discharge requirements as revisions to the Pennsylvania regulations.[2] According to the material released on that date, these revisions will go into effect upon publication in the *Pennsylvania Bulletin* as final rulemaking (as of June 30, that publication has not yet occurred).

Second, the U.S. Environmental Protection Agency (EPA) has adopted national discharge standards for many industries (known as effluent limitations guidelines or ELGs). The ELGs for the oil and gas industry are promulgated at 40 CFR Part 435 (Title 40, Part 435 of the *Code of Federal Regulations*). The ELGs specify zero discharge of produced water from onshore wells, but do allow two exceptions. The first applies to facilities located west of the 98th meridian (roughly the western half of the country) and the second applies to oil wells with very low production (less than 10 barrels of crude oil per day). Since all Marcellus Shale wells are located in the eastern United States and produce gas rather than oil, neither of these exceptions applies.

The May 17 rule revisions state that no discharge of oil and gas wastewater can be made directly from an oil and gas site to surface waters. Oil and gas wastewater can be sent to either a centralized treatment facility (referred to in this report as a commercial industrial wastewater treatment plant) or to a POTW. Those POTWs that accept wastewater from this category will be required to have an EPA-approved pretreatment program, which addresses TDS through local limits on these sources and at the above standards.

Reportedly, representatives of the West Virginia Department of Environmental Protection announced on May 19, 2010, that the agency would soon propose a new water quality standard for TDS of 500 mg/L.[3]

Current Practices from Surveyed Gas Companies: Argonne did not identify any Marcellus Shale gas producers that are directly discharging flowback or produced water from their well sites. There are several commercial wastewater disposal facilities that accept flowback and produced water, treat it, and discharge the water under their own NPDES permits. These facilities are discussed in a later section.

Haul to POTWs

Prior to the recent rapid development in the Marcellus Shale region, the oil and gas development activities in the region generated relatively small volumes of produced water. Some POTWs accepted limited quantities of produced water from oil and gas operators. The produced water was trucked from tanks or pits at the well site and discharged into the treatment facility. The treatment processes found at most POTWs are designed to remove suspended solids and biodegradable materials, but not salinity or TDS.

As the Marcellus Shale development grew in popularity, operators sought permission to bring more truckloads of salty flowback and produced water to the treatment plants. The increased input of TDS resulted in increased levels of TDS in the discharge. The May 17 revisions to the PADEP discharge regulations mentioned in the previous section will place restrictions on the volume of flowback and produced water that POTWs can accept.

Argonne obtained a list from PADEP of the POTWs and other commercial wastewater treatment facilities that currently accept or have applied for permits to accept flowback and produced water (Furlan 2010). A version of that table (edited to fit on a single-page width) is provided as Appendix A of this report. The table lists 15 POTWs that currently receive oil and gas water or have received it in the past. Many of those POTWs have conditions in their NPDES permits requiring that the volume of wastewater from oil and gas sources may not exceed 1% of the average daily flow. The table lists 4 other POTWs that receive treated water from commercial wastewater treatment companies that discharge to the municipal sewer system rather than discharging directly into surface water bodies.

Current Practices from Surveyed Gas Companies: Three of the surveyed gas companies are currently sending flowback and produced water to POTWs or have done so in the past. One of the replies indicated that it had sent its flowback and produced water to the New Castle, Pennsylvania, wastewater treatment plant. The other two replies did not identify the municipality receiving the water.

Haul to Commercial Industrial Wastewater Treatment Plant

Several Pennsylvania companies have provided wastewater disposal service to the oil and gas community for many years. As the volume of flowback and produced water has increased rapidly over the past few years, new commercial disposal companies are opening their doors, while still others are applying for permits from PADEP.

Argonne obtained a list of commercial industrial wastewater treatment plants that currently accept flowback and produced water from the PADEP (Furlan 2010). This list is part of the large table provided in Appendix A. Twenty-seven commercial wastewater treatment facilities (noted as CWT in Appendix A) are permitted by the PADEP to treat flowback and produced water and then discharge the treated water to surface water bodies. Four other commercial facilities treat the water and then discharge it to municipal sewers that flow to POTWs. The PADEP list also includes 25 other commercial wastewater treatment facilities that have applied for permits but have not yet received permission to operate and discharge.

Additional information about the commercial wastewater treatment industry in Pennsylvania was obtained from an earlier Argonne report. Puder and Veil (2006) report that eight facilities in Pennsylvania were accepting produced water in the time period between October 2005 and April 2006. However, only four of them provided information that could be used in that study. Table 2 shows the Pennsylvania facilities that were listed in Puder and Veil (2006). The costs listed in Table 2 represent the costs during 2005-2006 and may not be representative of costs now.

These four companies are still providing wastewater disposal service to the oil and gas industry. The May 17 revisions to the PADEP discharge regulations include an important provision relating to existing commercial industrial disposal companies. Any commercial industrial disposal company with a valid NPDES permit as of the date on which the rule revisions are finalized (presumably during the summer of 2010) is allowed to continue discharging at the permitted levels until such time as the facility seeks an increase in discharge allowance. This provision was included in the rule revisions to allow the existing disposal capacity to remain in place. Each facility has permit limits that were calculated to allow TDS discharges without violating surface water quality.

Table 2. Commercial Disposal Facilities in Pennsylvania that Accepted Produced Water in the 2005-2006 Time Frame (based on Puder and Veil 2006)

Disposal Facility Name	Location	Disposal Cost	Comments	Throughput Capacity*
Castle Environmental Inc. (name changed to Advanced Waste Services of Pennsylvania in 2010)	New Castle, PA	$0.025– $0.050/gal	The facility operates a nonhazardous wastewater processing facility. Treatment involves chemical precipitation and filtration. The resulting water from the process is discharged to the New Castle Sanitation Authority's Wastewater Treatment Plant.	Daily volume on 6/28/10 was ~260,000 gal/day; this was slightly lower than average
Hart Resource Technologies	Creekside, PA	$0.0525/gal	Treatment involves chemical precipitation and removal of oils and heavy metals. Surface water discharge occurs under an NPDES permit issued by PADEP.	18,000 gal/day of produced water; 45,000 gal/day of flowback; total from 11/08 to 10/09 = 23.2 million gal
Pennsylvania Brine Treatment	Franklin, PA	$ 0.055/gal	Facility uses chemical precipitation and generates nonhazardous residual sludge that is land-filled offsite at a PADEP-permitted facility. The treated water is then discharged to surface waters under an NPDES permit.	140 gpm; total from 11/08 to 10/09 = 53 million gal
Tunnelton Liquids	Saltsburg, PA	$0.045/gal	Facility uses an innovative process to treat pit water (containing some oil- based muds and cuttings). It combines acid mine drainage from an abandoned coal mine with the produced water. Sulfates in the mine drainage help remove contaminants from the produced water. Following several treatment steps, the treated water is discharged to a river under the authority of an NPDES permit.	Total of 1 million gal/day; ~100,000 gal/day of oil and gas water; ~900,000 gal/day of acid mine drainage

* The estimate for Castle Environmental was provided by phone on June 29, 2010 (Meahl 2010). The other estimates were provided during the May 2010 site visits described in Chapter 4.

To get a first-hand look at how some of these facilities operate, the author visited four facilities located in different parts of Pennsylvania during May 2010. Those site visits are described in Chapter 4.

Current Practices from Surveyed Gas Companies: Five of the companies noted that they are currently sending some of their flowback and produced water to commercial disposal companies. Most of the disposal companies are located in Pennsylvania, but one was located in West Virginia.

Reuse for a Future Frac Job

The gas companies are interested in finding water to use in frac jobs and in managing the subsequent flowback and produced water from those wells in ways that minimize costs and environmental impacts. One way to accomplish this goal is to collect the flowback water and reuse it for frac fluids in other wells. Several gas companies are currently using this approach.

The May 17 revisions to the PADEP discharge regulations also include a requirement that any oil and gas wastewater having TDS of less than 30,000 mg/L cannot be discharged but must be recycled and reused.

The chemical composition of frac fluids is designed to optimize the performance of the frac job. Generally, the TDS concentration of the flowback and produced water is higher than the desired TDS range for new frac fluids. Several Marcellus Shale operators start with flowback and produced water and blend it with enough freshwater from some other source to reduce TDS and other constituents to fall within an acceptable concentration range. At least one other Marcellus Shale operator is using a thermal distillation process to treat the salty flowback and produced water to make very clean water. After passing through the treatment unit (known as the AltelaRain® system — described in Veil [2008]), the water is segregated into a freshwater stream and a concentrated brine stream. The concentrated brine stream is hauled offsite for disposal at a commercial disposal facility while the clean water can be reused for a future frac job or possibly some other use.

Other technology providers are attempting to find niches in the Marcellus Shale water treatment market. For example, several Aqua-Pure thermal distillation units (described in Veil [2008]) were licensed in 2009 by Eureka Resources, a commercial wastewater treatment facility in northern Pennsylvania, to provide a high degree of TDS removal when needed. Another flowback and produced water treatment operation was opened for business in 2009 in West Virginia by AOP Clearwater. Its website[4] does not provide any details about the actual treatment process, but it does indicate that the process results in "distilled water," suggesting that some type of thermal distillation process is used.

In April 2010, the PADEP issued general permit WMGR121 for Processing and Beneficial Use of Gas Well Wastewater from Hydraulic Fracturing and Extraction of Natural Gas from the Marcellus Shale Geological Formation.[5] The PADEP list included in Appendix A shows two facilities that operate under authorization by the general permit.

One of the projects funded by the DOE/NETL in September 2009 will assemble a trailer containing various types of flowback and produced water treatment devices. The trailer will be deployed to several Marcellus Shale locations for tests on the flowback and produced water from gas wells. The project is headed by Texas A&M University. This research, when completed, will contribute additional data and experience with other types of treatment processes.

Current Practices from Surveyed Gas Companies: Six of the seven companies indicated that they are reusing at least some of their flowback and produced water for future frac jobs. Several of these companies are attempting to recycle all of the flowback and produced water they generate. The seventh company is operating only a single Marcellus Shale well at this time. That company noted that it plans to recycle its flowback and produced water in the future as it develops more wells.

Results from Operator Survey

The options actually used by Marcellus Shale gas operators were briefly mentioned at the end of each of the previous sections. In this section, more detailed information is attributed to particular operators and is compiled into a single table (Table 3).

To summarize, many options are currently being employed to manage flowback and produced water. Most of the operators are recycling some, to all, of their flowback and produced water. The flowback and produced water that is not being recycled is hauled offsite to POTWs, commercial wastewater disposal facilities, or commercial injection wells.

Table 3. Water Management Options Used by Selected Marcellus Shale Operators

Company	Information Provided by the Company Contact
Chesapeake Energy	Chesapeake Energy manages its flowback and produced water generated in accordance with local, state, and Federal regulations applicable in each state within the Marcellus Play. It conducts ongoing research to identify environmentally safer methods of byproduct management. At various locations, Chesapeake has transported flowback and produced water offsite to a commercial wastewater disposal company, transported it offsite to a sewage treatment plant, and has treated the water for reuse. Injection wells are also used where practical; however these wells are not on the same location as the producing well (Gillespie 2010).
Range Resources	Range Resources is trying to reuse 100% of its flowback, production brine, and drill pit water. The only "processes" involved are settling and dilution. Working backwards from the well performance it sees that it gets just as good of a result with diluted reuse water as with all freshwater. Range Resources has no indication of issues with frac fluid stability, scaling, or bacteria growth downhole. It uses conventional slickwater fracturing additives and designs its fracs based upon the final diluted fluid (Gaudlip 2010a). During 2009, Range Resources completed 44 wells and did frac jobs involving 364 stages. The total volume of frac fluid used was 158 million gallons, with 28% of the volume made up of recycled water from a previous well. The estimated cost savings from avoided disposal fees, less freshwater purchased, and less trucking costs was $3.2 million. The wells that included recycled water accounted for 17% of Range Resources' Marcellus wells. Fifty percent of the wells that used recycled water are in the company's top 25 producing wells (Gaudlip 2010b).
EQT	EQT reuses all of its flowback water without treating it. Flowback water is trucked to the next well location where it is blended with freshwater. Some of the ongoing produced water is hauled to a commercial disposal facility in West Virginia (AOP Clearwater), while other produced water is hauled to a commercial disposal well in Ohio (Babich 2010).

Table 3. (Continued)

Company	Information Provided by the Company Contact
East Resources	East Resources recycles all of its produced water and drilling pit fluids into frac fluids used in other wells. The produced water is not treated for TDS but is blended with freshwater. A typical well uses 3.5 million gallons of water in frac fluids. East Resources generally gets 18% to 20% of the water back to the surface. The company is looking at alternate sources of freshwater such as mine water, produced water from shallow formations, and treated POTW effluent (Blauvelt 2010).
BLX	BLX is a small producer and does not drill as many wells as some larger companies. Therefore, the treatment of water for reuse does not work for them because of the length of time between frac jobs. BLX has hauled water to another site, if available, and basically diluted it with freshwater. The rest of the water goes to a disposal well, a sewage plant in New Castle, Pennsylvania, or one of the offsite commercial disposal facilities (Berdell 2010). BLX is involved with at least two wells that are using the AltelaRain® thermal distillation technology to treat produced water. One of these wells is described in a presentation made to the Pennsylvania Senate (Kohl 2010). The other well is part of a DOE/NETL funded project that will characterize the results of the AltelaRain treatment process when used to treat flowback and produced water from a Marcellus Shale well.
Norse Energy	Norse Energy operates in New York and currently has only one operating Marcellus well. Additional wells have been proposed, but are being delayed, awaiting approval from the New York Department of Environmental Control. For its existing Marcellus well, Norse has disposed of the fluid at two facilities located in Warren, Pennsylvania and Franklin, Pennsylvania. Norse is looking at additional disposal sites in the Williamsport, Pennsylvania, area for future wells. Regardless of the current facility Norse is using, trucking costs are a major portion of the total disposal cost. Norse would also consider transporting the flowback and produced water to a sewage treatment plant, if available, in order to reduce transportation costs (Keyes 2010). Norse Energy would consider several other options. If injection were possible, Norse would most likely use a centralized injection well with a water-gathering system from several producing wells or pads. Norse is also looking at treating and reusing the water in future wells (Keyes 2010).

Site Visits to Commercial Wastewater Disposal Facilities

During May 2010, the author visited four commercial wastewater disposal facilities in Pennsylvania. All of these facilities accept flowback and produced water from Marcellus Shale gas production. Figure 17 shows the location of each facility.

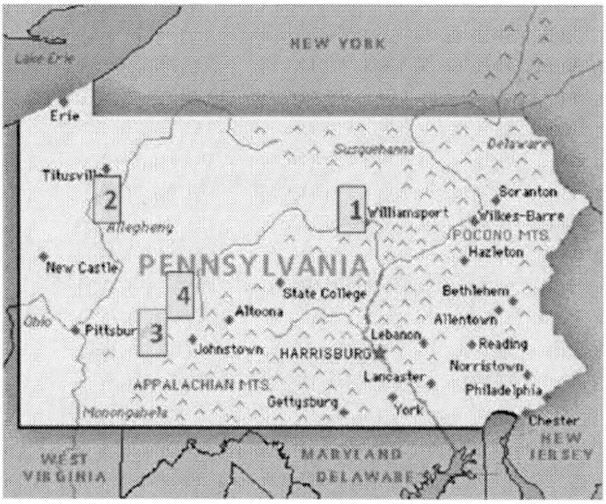

Figure 17. Locations of Commercial Wastewater Treatment Facilities Visited.

Eureka Resources

Eureka Resources operates a commercial wastewater treatment facility in Williamsport, Pennsylvania. Dan Ertel provided a tour of the facility on May 10. Figures 18 through 23 are photos taken during the tour. Trucks unload into one of four settling tanks to allow settling of heavy solids and removal of any free oil. The water is sent to treatment tanks where the pH is raised using sodium sulfate (Na_2SO_4) or lime to facilitate the removal of dissolved barium and other metals. Coagulants are added to aid in settling; then the water flows to clarifiers to settle. Solids are dewatered in a type of filter press called a "membrane squeeze press." The treated water is discharged to the Williamsport municipal sewer system and sent to the city's wastewater treatment plant. The current process removes metals, but it does not remove TDS. The throughput of the plant is 300,000 gallons per day.

Eureka Resources is in the process of installing two Aquapure NOMAD evaporators that are expected to go into service during June 2010. The NOMAD units (Veil 2008) will remove nearly all of the TDS from the water. The concentrated brine solution produced by the NOMAD units will be either transported to a disposal well in Ohio or sent to a local company that will evaporate the brine to produce salt.

Figure 18. Unloading Area - Eureka Resources.

Figure 19. Treatment Tanks – Eureka Resources.

Figure 20. Additional Treatment Tank – Eureka Resources.

Figure 21. Filter Press – Eureka Resources.

Figure 22. View of Part of NOMAD Unit – Eureka Resources.

Figure 23. Additional View of NOMAD Unit – Eureka Resources.

Pennsylvania Brine

Pennsylvania Brine operates two commercial wastewater treatment facilities in western Pennsylvania (in the towns of Franklin and Josephine). Elton Delong provided a tour of the Franklin facility on May 10. Figures 24 through 27 are photos taken during the tour. The facility was not processing any water that day as it was undergoing renovations. Mr. Delong described the process units that will be in place once the renovations are completed.

Figure 24. Unloading Area – Pennsylvania Brine.

Trucks unload on a pad. The flowback and produced water is screened to remove large objects, then flows into a settling tank to allow settling of heavy solids and removal of any free oil. The water continues to an aeration tank, then moves to another tank where lime is added. Following the renovations, Na_2SO_4 will be added to the water, prior to moving it to the lime tank. Next, the water flows to a tank where polymers are added to promote coagulation; then it moves to a clarifier to settle. Solids are dewatered in a filter press. Acid is added to the treated water to return the pH to a neutral range. Finally, the treated water is discharged to the Allegheny River under an NPDES permit issued by the PADEP. The current throughput of the plant is about 140 gallons per minute. According to data supplied by the company, in the one-year period from November 2008 to October 2009, the facility processed about 53 million gallons of water. The process removes metals, but does not remove TDS.

Figure 25. Settling Tank – Pennsylvania Brine.

Figure 26. Treatment Tank – Pennsylvania Brine.

Figure 27. Clarifier – Pennsylvania Brine.

Tunnelton Liquids

Tunnelton Liquids operates a commercial wastewater treatment facility in Saltsburg, Pennsylvania. The facility was originally constructed to treat acid mine drainage from a coal refuse pile located uphill from the treatment plant. Bruce Bufalini provided a tour of the facility on May 14. Figures 28 through 33 are photos taken during the tour.

Trucks unload at two different pads. Flowback and produced water are sent to an oil/water separator first to remove any free oil. The separator effluent flows to a large pond called the "raw pond." Water from onsite pits flows directly to the raw pond as does leachate from the coal refuse pile. Water is pumped from the raw pond to a treatment plant. The first stage at the plant is an aeration tank where lime is added. If the aeration creates excessive foaming, an anti-foaming chemical is added to the tank. The water then flows to a clarifier. The collected solids are removed periodically and they are disposed of in a deep mine where they help to neutralize the acidic mine pool water. The treated water flows to a final polishing basin that has several days of retention time. Finally, the treated water is discharged to the Conemaugh River under an NPDES permit issued by the PADEP.

The plant has an average discharge of about one million gallons per day. Only about 100,000 gallons are oil and gas flowback and produced water — the rest is acid mine drainage. The process removes metals, but does not remove TDS.

Figure 28. Unloading Area – Tunnelton Liquids.

Figure 29. Oil/Water Separator – Tunnelton Liquids.

Figure 30. Raw Pond – Tunnelton Liquids.

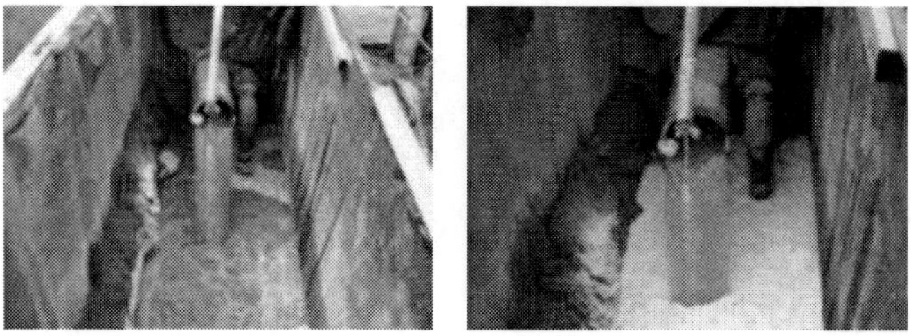

Figure 31. Aeration Tank – with and without Anti-Foaming Chemical – Tunnelton Liquids.

Figure 32. Clarifier – Tunnelton Liquids.

Figure 33. Polishing Pond (visible behind trees) – Tunnelton Liquids.

Hart Resource Technologies

Hart Resource Technologies operates a commercial wastewater treatment facility in Creekside, Pennsylvania. Like the Tunnelton facility, the Hart facility was originally constructed to treat acid mine drainage. Paul Hart and Becky Snyder provided a tour of the facility on May 14. Figures 34 through 37 are photos taken during the tour. The plant operates two separate, but equivalent, treatment systems. One system treats "low-salt" water (defined as lower than about 40,000 mg/L TDS) and the other system treats "high-salt" water.

From the unloading pad, the flowback and produced water passes through a strainer to remove large debris. It then flows to a settling pit, followed by an oil/water separator. If any of the water contains oil/water emulsions, it is further treated by some combination of heat, demulsifying chemical, and increased retention time. Water next moves into a storage tank where blending of water from several truckloads occurs. From here, the water is treated in batches in a treatment tank. Several sequential steps occur in the treatment tank (agitation, aeration, and pH adjustment with lime). The low-salt treatment system receives a flocculant, while the high-salt system receives Na_2SO_4. Following these steps, the water is allowed to settle in the same tank. The water then flows to a clarifier. The sludge from the settling processes and the clarifier are treated in a thickener and then dewatered in a filter press. Solids are sent to a local landfill.

Finally, the treated water is discharged to McKee Run under an NPDES permit issued by the PADEP. The pH is not neutralized — the high pH in the discharge helps to neutralize the low instream pH caused by local acid mine drainage. The plant has an average discharge of about 18,000 gallons per day of produced water and 45,000 gallons per day of flowback water. The process removes metals but does not remove TDS. Mr. Hart and Ms. Snyder indicated that the facility processed more water from shallow gas wells in the region than water from Marcellus Shale wells.

Figure 34. Unloading Area, Settling Pit, and Blending Tank – Hart Resource Technologies.

Figure 35. Treatment Tank – Hart Resource Technologies.

Figure 36. Clarifier – Hart Resource Technologies.

Figure 37. Filter Press for Solids – Hart Resource Technologies.

FINDINGS AND CONCLUSIONS

Findings

There is a great deal of interest in natural gas production in the Marcellus Shale. The Marcellus offers hope of substantial energy resources and economic benefits, but also creates various environmental and societal issues. This report describes three types of water issues that arise from shale gas development in the Marcellus Shale. Those three issues are:

- Controlling the stormwater runoff from disturbed areas,
- Obtaining sufficient freshwater supply to conduct frac jobs on new wells, and
- Managing the flowback water and produced water from the well.

In particular, the report focuses on the third of these issues. Some of the key findings are listed below:

1) Marcellus Shale frac jobs typically inject several million gallons of frac fluids (which are mostly water). In some of the other shale gas plays around the country, the flowback volume can be 30% to 70% of the initial volume of injected fluids. But in the Marcellus, the flowback volume appears to be lower than that — often lower than 25%.
2) After the initial return of flowback water, within a few weeks following completion of the frac job, most wells continue generating formation water (produced water) at a lower rate for many years.
3) Historically, oil and gas operators in Pennsylvania transported their produced water offsite to disposal wells in Ohio, POTWs, or commercial industrial disposal facilities. These practices continue today. However, the increased volume of flowback and produced water from the expanding Marcellus shale gas industry has taxed the capacity of facilities to manage the water and has occasionally resulted in elevated levels of TDS in some of Pennsylvania's rivers and streams. As a result, more treatment facilities are coming on-line or are being permitted.
4) In May 2010, Pennsylvania adopted new, more-stringent discharge requirements for oil and gas flowback and produced waters. Commercial disposal companies that already hold discharge permits are grandfathered to discharge at their current levels. New dischargers face much more restrictive limits on TDS.

CONCLUSIONS

Gas production in the Marcellus Shale region is expanding rapidly. State agencies face new challenges in managing and regulating a growing number of wells. New policies and regulations continue to evolve (e.g., Pennsylvania's May 2010 revisions to discharge regulations for oil and gas wastewater). This report examines the available flowback and water management technologies and methods used today and likely to be used in the next few years. Some of the conclusions that follow from the report include:

1) Unlike shale gas plays in arid states (e.g., the Barnett Shale play in Texas), the Marcellus Shale occurs in a part of the country that generally has sufficient water supplies. Obtaining water for frac jobs, while necessitating coordination with various agencies, including the

SRBC and the Delaware River Basin Commission (DRBC), has not yet proven to be a barrier. If the number of new shale gas wells continues to rise rapidly, water supplies could become a barrier.

2) Several of the Marcellus gas operators have begun recycling their flowback water into new frac fluids. They have experimented with mixing the high-TDS flowback with lowTDS freshwater to make intermediate-TDS frac water. The early results from this work seem promising. If recycling can be practiced more widely throughout the region, companies can save costs on disposal fees and trucking fees, while reducing the volume of freshwater used for new frac fluids.
3) The regulatory environment is contentious and evolving. Opponents of gas drilling, landowners hoping to gain substantial income from leasing mineral rights, gas companies, and politicians seeking jobs for their constituents will continue to debate how and where gas should be produced within the Marcellus region.

References

Adams, R., 2010, email communication from Richard Adams, Chief Oil and Gas, to John Veil, Argonne National Laboratory, March 30.

Babich, M.A., 2010, email communication from Mary Anna Babich, EQT-Production, to John Veil, Argonne National Laboratory, and a follow-up phone conversation between Babich and Veil, both on March 30.

Berdell, S., 2010, email communication from Stan Berdell, BLX, to John Veil, Argonne National Laboratory, April 23.

Blauvelt, S., 2010, telephone conversation between Scott Blauvelt, East Resources, and John Veil, Argonne National Laboratory, April 6.

Furlan, R., 2010, email communication including a spreadsheet of wastewater disposal sites, from Ronald Furlan, PADEP, to John Veil, Argonne National Laboratory, June 30.

Gaudlip, T., 2010a, email communication from Tony Gaudlip, Range Resources, to John Veil, Argonne National Laboratory, March 12.

Gaudlip, T., 2010b, "Preliminary Assessment of Marcellus Water Reuse," presented at Process-Affected Water Management Strategies conference, Calgary, Alberta, Canada, March 17.

Gillespie, E., 2010, email communication from Eric Gillespie, Chesapeake Resources, to John Veil, Argonne National Laboratory, March 31.

GWPC and ALL, 2009, "Modern Shale Gas Development in the United States: A Primer," prepared by the Ground Water Protection Council and ALL Consulting for the U.S. Department of Energy, National Energy Technology Laboratory, April, 116 pp.

Hoffman, J., 2010, "Susquehanna River Basin Commission Natural Gas Development," presented at the Science of the Marcellus Shale Symposium at Lycoming College, Williamsport, PA, January 29. Available at http://www.srbc.net/programs/projreviewmarcellustier3.htm; accessed April 27, 2010.

Keyes, S., 2010, email communication from Steve Keyes, Norse Energy, to John Veil, Argonne National Laboratory, March 31.

Kohl, D., 2010, presentation to Pennsylvania Senate Environmental Resources and Energy Committee, January 27. Available at http://www.senatormjwhite.com/environmental/2010/012710/kohl.pdf; accessed May 3, 2010.

Meahl, P., 2010, telephone conversation between Paul Meahl, Advanced Waste Service of Pennsylvania, and John Veil, Argonne National Laboratory, June 29.

Miller, G., 2010, email communication including a spreadsheet of injection well volumes from Gregg Miller, Ohio DMRM, to John Veil, Argonne National Laboratory, June 30.

Puder, M.G., and J.A. Veil, 2006, *Offsite Commercial Disposal of Oil and Gas Exploration and Production Waste: Availability, Options, and Costs*, ANL/EVS/R-06/5, prepared by the Environmental Science Division, Argonne National Laboratory for the U.S. Department of Energy, Office of Fossil Energy and National Energy Technology Laboratory, August. Available at http://www.evs.anl.gov/pub/dsp_detail.cfm?PubID=2006; accessed April 28, 2010.

Tomastik, 2010, email communication from Tom Tomastik, Ohio DMRM, to John Veil, Argonne National Laboratory, April 26.

Veil, J.A., 2008, *Thermal Distillation Technology for Management of Produced Water and Frac Flowback Water*, Water Technology Brief #2008-1, prepared for U.S. Department of Energy, National Energy Technology Laboratory, May 13, 12 pp. Available at http://www.evs.anl.gov/pub/dsp_detail.cfm?PubID=2321; accessed May 3, 2010.

Appendix A: Pennsylvania Facilities Permitted to Accept Oil and Gas Wastewaters and Other Facilities that Have Applied for Permits to Accept Oil and Gas Wastewater

Facility Name	Permit Number	Receiving Stream	Design Flow (MGD)	Oil & Gas Wastewater Flow (MGD)	County	TYPE[a]	Status / NOTES	Facility Designation[b]
Clairton Municipal Authority	PA0026824	Peters Creek	6.0	0.035	Allegheny	POTW	O&G waste flow limited to 1% of Avg Daily flow at POTW by DEP order.	O&G, MSW
Municipal Authority City of McKeesport	PA0026913	Monongahela River	11.5	0.102	Allegheny	POTW	O&G waste flow limited to 1% of Avg Daily flow at POTW by DEP order.	O&G, MSW
Allegheny Valley Joint Sanitary Authority	PA0026255	Allegheny River	5.5	0.025	Allegheny	POTW	Limiting O&G waste flow to 25,000 gpd and chlorides to 24,000 mg/l	O&G, MSW
CNX Gas Co LLC	PA0253588	Crooked Creek	0.15	0.15	Armstrong	CWT	n/a	CBM
Somerset Regional Water Resources	PA0233901	North Branch Susquehanna	0.5	0.5	Bradford	Proposed CWT	Metals precipitation, filtration (reverse osmosis), and thermal evaporation for RO	PMSW

Appendix A. (Continued)

Facility Name	Permit Number	Receiving Stream	Design Flow (MGD)	Oil & Gas Wastewater Flow (MGD)	County	TYPE[a]	Status / NOTES	Facility Designation[b]
Vavco LLC	Approval 1009001 under GP PAG310001	Little Connoquenessing Creek	0.0009	0.0009	Butler	Proposed CWT	Proposed facility to treat only oil well production fluids from own stripper oil wells. Approval issued 7/1/2009. Part II WQM application received 3/31/2010. Technical deficiency letter sent May 27, 2010.	SW, PO&G
Johnstown Redevelopment Authority - Dornick Point STP	PA0026034	Conemaugh River	12.0	0.076	Cambria	POTW	Limiting O&G waste flow to 1% of Avg Daily flow.	O&G, MSW
Great Lakes Energy Partners, LLC	PA0253103	Clearfield Creek	0.2	0.2	Cambria	CWT	n/a	CBM

Facility Name	Permit Number	Receiving Stream	Design Flow (MGD)	Oil & Gas Wastewater Flow (MGD)	County	TYPEa	Status / NOTES	Facility Designationb
Keystone Clearwater Solutions, LLC	PA0233951	Moshannon Creek	0.504	0.504	Centre	Proposed CWT	Metals precipitation and filtration treatment (nanofiltration and RO). Had some land use issues that are now resolved. This application was returned, but has been resubmitted with appropriate treatment proposal.	PMSW
Clearfield Municipal Authority	PA0026310	West Branch Susquehanna River	4.5	0.01	Clearfield	POTW	This facility has claimed to have historically taken O&G wastewater. Submitted an amended NPDES application to take wastewater until December 31, 2010.	O&G
Dannic Energy Corp.	PA0233790	Hawk Run	0.25	0.25	Clearfield	Proposed CWT	Metals precipitation and vacuum evaporator for treatment.	PMSW

Appendix A. (Continued)

Facility Name	Permit Number	Receiving Stream	Design Flow (MGD)	Oil & Gas Wastewater Flow (MGD)	County	TYPE[a]	Status / NOTES	Facility Designation[b]
Keystone Clearwater Solutions, LLC	PA0233960	West Branch Susquehanna River	0.504	0.504	Clinton	Proposed CWT	Metals precipitation and filtration treatment (nanofiltration and RO). This application was returned, but has been resubmitted with appropriate treatment proposal.	PMSW
Dannic Energy Corp.	PA0233781	West Branch Susquehanna River	0.25	0.25	Clinton	Proposed CWT	Hyner Drilling Fluid Recycling Facility. Metals precipitation and vacuum evaporator for treatment.	PMSW
Central PA Wastewater, Inc.	PA0233706	Unnamed tributary to West Branch Susquehanna River	0.4	0.4	Clinton	Proposed CWT	Discharge is proposed to intermittent stream less than 0.5 miles from the river.	PMSW

Facility Name	Permit Number	Receiving Stream	Design Flow (MGD)	Oil & Gas Wastewater Flow (MGD)	County	TYPE[a]	Status / NOTES	Facility Designation[b]
							Metals precipitation is the proposed treatment technology. Application needs to be updated to show advanced treatment to meet stringent TDS limits.	
Ridgway Borough	PA0023213	Clarion River	2.2	0.02	Elk	POTW	n/a	O&G, MSW
ProChemTech International-Blue Valley Hydrofrac Recycle Facility	PA0268305	Unnamed tributary to Brandy Camp Creek	0.0125	0.3	Elk	Proposed CWT	Proposed permit to accept 300,000 gallons per day of Marcellus Shale wastewater for treatment and recycle. Proposing to discharge a maximum of 12,500 gallons per day of distilled water.	PMSW

Appendix A. (Continued)

Facility Name	Permit Number	Receiving Stream	Design Flow (MGD)	Oil & Gas Wastewater Flow (MGD)	County	TYPE[a]	Status / NOTES	Facility Designation[b]
Veolia ES Greentree Landfill LLC	PA0103446	Little Toby Creek	0.25	0.04808	Elk	CWT	Facility treats 48,000 gallons of natural gas well brine water per year from American Refining and Exploration, Inc. and 60-80 gallons of natural gas well brine water per year from Destiny, Inc. - both non Marcellus natural gas wells.	O&G
Shallenberger Construction – Ronco Facility	PA0253723	Monongahela River	0.5	0.5	Fayette	CWT	Part II permit issued for Phase I of the treatment plant. Consent Order and Agreement (COA) includes compliance schedule. Water Quality Management Part II permit was issued on August 28, 2009 for the construction of Phase I of this treatment facility, which does not include treatment units for TDS.	MSW, O&G

Facility Name	Permit Number	Receiving Stream	Design Flow (MGD)	Oil & Gas Wastewater Flow (MGD)	County	TYPEa	Status / NOTES	Facility Designationb
							An appeal of the issuance of the Part II permit and the COA was filed on behalf of Clean Water Action. The draft NPDES permit amendment, incorporating TDS and other effluent limitations was sent for publication. Facility is currently in operation, with treated wastewater being hauled to the next drill site.	
Shallenberger Construction	PA0253863	Youghiogheny River	1.0	1.0	Fayette	Proposed CWT	Oil and Gas wastewaters-CWT— Proposed Facility	PMSW, PO&G
Municipal Authority of Belle Vernon	PA0092355 and PA0092355 -A1	Monongahela River	0.5	0.0	Fayette	POTW	O&G waste flow limited to 1% of Avg Daily flow at POTW by DEP order. This facility no longer accepts large volumes of O&G wastewater.	O&G, MSW

Appendix A. (Continued)

Facility Name	Permit Number	Receiving Stream	Design Flow (MGD)	Oil & Gas Wastewater Flow (MGD)	County	TYPE[a]	Status / NOTES	Facility Designation[b]
Brownsville Municipal Authority	PA0022306	Dunlap Creek	0.96	0.0	Fayette	POTW	O&G waste flow limited to 1% of Avg Daily flow at POTW by DEP order. This facility no longer accepts large volumes of O&G wastewater.	O&G, MSW
Green Earth Wastewater	PA0253821	Dunkard Creek	0.25	0.25	Greene	Proposed CWT	O&G wastewaters- CWT— Proposed Facility	PMSW, PO&G
Waynesburg Borough	PA0020613	South Fork Tenmile Creek	0.8	0.0	Greene	POTW	O&G waste flow limited to 1% of Avg Daily flow at POTW by DEP order. This facility no longer accepts large volumes of O&G wastewater.	O&G, MSW
Franklin Township Sewer Authority/Tri-County Wastes (CWT)	PA0046426	South Fork Tenmile Creek	1.25	0.05	Greene	POTW receiving indirect discharge from CWT	O&G waste flow will be limited to 50,000 gpd by DEP order. Waste is pretreated but not for TDS or Chlorides.	O&G, MSW
CNX Gas Co LLC	PA0252611	Blockhouse Run	0.007	0.007	Greene	CWT	n/a	CBM

Facility Name	Permit Number	Receiving Stream	Design Flow (MGD)	Oil & Gas Wastewater Flow (MGD)	County	TYPE[a]	Status / NOTES	Facility Designation[b]
CNX Gas Co LLC	PA0252832	Pennsylvania Fork Fish Creek	0.0111	0.0111	Greene	CWT	n/a	CBM
CNX Gas Co LLC – Rogersville Treatment Facility	PA0253286	South Fork Tenmile Creek	0.202	0.202	Greene	Proposed CWT	CNX proposes to construct the Rogersville Treatment Facility to treat waters associated with natural gas production at its Greenhill Production Area. The proposed plant will be able to treat 201,600 gpd. The site will have two mobile reverse osmosis treatment plants on site along with 12 concentrate storage tanks and one clean water impoundment. This is a recycle and reuse plant.	PMSW, PO&G

Appendix A. (Continued)

Facility Name	Permit Number	Receiving Stream	Design Flow (MGD)	Oil & Gas Wastewater Flow (MGD)	County	TYPE[a]	Status / NOTES	Facility Designation[b]
							The applicant has not submitted an NPDES permit application for the overflows from the impoundment. The application does not meet the residual waste regulations. A technical deficiency letter will be sent to the applicant.	
PA Brine Josephine (Franklin Brine)	PA0095273	Blacklick Creek	0.12	0.12	Indiana	CWT	O&G wastewaters. Currently under detailed compliance review.	O&G, MSW
Hart Resource	PA0095443	McKee Run	0.045	0.045	Indiana	CWT	O&G wastewaters. Currently under detailed compliance review.	O&G, MSW
Tunnelton Liquids	PA0091472	Conemaugh River	1.0	1.0	Indiana	CWT	n/a	O&G, MSW

Facility Name	Permit Number	Receiving Stream	Design Flow (MGD)	Oil & Gas Wastewater Flow (MGD)	County	TYPE[a]	Status / NOTES	Facility Designation[b]
Frontier Energy Services	PA0254207	Yellow Creek	0.9	0.9	Indiana	Proposed CWT	Proposed 0.9 mgd CWT for drilling wastes. Applicant proposed to treat wastes through reverse osmosis and evaporation.	PMSW, PO&G
Belden and Blake	PA0219339	Blacklick Creek	0.6	0.6	Indiana	CWT	n/a	CBM
CNX Gas Co LLC	PA0253596	Blacklegs Creek	0.035	0.035	Indiana	CWT	n/a	CBM
Canton Oil & Gas Company	PA0206075	Blacklick Creek	0.48	0.48	Indiana	CWT	n/a	CBM
CNX Gas Co LLC	PA0253995	Kiskiminetas River	0.21	0.21	Indiana	Proposed CWT	Application currently under review process.	PMSW, PO&G
Punxsutawney Borough	PA0020346	Mahoning Creek	2.2	0.02	Jefferson	POTW		O&G

Appendix A. (Continued)

Facility Name	Permit Number	Receiving Stream	Design Flow (MGD)	Oil & Gas Wastewater Flow (MGD)	County	TYPE[a]	Status / NOTES	Facility Designation[b]
Brockway Area Sewage Authority	PA0028428	Toby Creek	1.5	0.014	Jefferson	POTW	Currently 14,000 GPD to POTW. Request to increase brine volume withdrawn. NPDES renewal drafted 6/19/09. EPA issued a general objection letter and 90-day time extension on July 17, 2009.	O&G
Reynoldsville Boro	PA0028207	Sandy Lick Creek	0.8	0.011	Jefferson	POTW	Have taken in approx. 14,000 gpd of brine to POTW for a long time.	O&G
Dominion Transmission Corp - DivV	PA0101656	Stump Creek	0.01008	0.0077	Jefferson	CWT	Facility only treats Dominion's own wastewater. Most likely no Marcellus Shale water.	O&G
New Castle City / Advanced Waste Services	PA0027511	Mahoning River	17	0.55	Lawrence	POTW receiving discharge from CWT	The CWT, Advanced Waste Services discharges O&G wastewater to New Castle POTW. Possible expansion plans from 0.2 MGD to 2.2 MGD	O&G, MSW

Facility Name	Permit Number	Receiving Stream	Design Flow (MGD)	Oil & Gas Wastewater Flow (MGD)	County	TYPE[a]	Status / NOTES	Facility Designation[b]
Williamsport Sanitary Authority / Eureka Resources	PA0027057	West Branch Susquehanna River	8.4	0.12	Lycoming	POTW receiving discharge from CWT	Town has an EPA approved pretreatment program. Installing mechanical vapor recompression evaporation for treatment summer 2010.	MSW
Water Treatment Solutions	PA0233838	Daugherty's Run	0.05	0.05	Lycoming	Proposed CWT	Metal precipitation and thermal distillation/crystallizer treatment; limited information submitted.	PMSW
Dannic Energy Corp.	PA0233765	Pine Run	0.25	0.25	Lycoming	Proposed CWT	Pine Run discharges to W. Branch Susquehanna. Metals precipitation and vacuum evaporator for treatment.	PMSW
TerrAqua Resource Management	A0233650	West Branch Susquehanna River	0.4	0.4	Lycoming	Proposed CWT	Metals precipitation and thermal distillation is the proposed treatment technology. Part I NPDES permit issued; awaiting Part II WQM application	PMSW

Appendix A. (Continued)

Facility Name	Permit Number	Receiving Stream	Design Flow (MGD)	Oil & Gas Wastewater Flow (MGD)	County	TYPE[a]	Status / NOTES	Facility Designation[b]
TerrAqua Resource Management LLC	WMGR121	None	0.4	0.4	Lycoming	RW Processing	Batch chemical processing - treated wastewater returned to well sites for reuse	MSW
Minard Run Oil Company	PA0105295	Lewis Run	0.016	0.004	McKean	CWT	Existing facility which treats only oil well production fluids from own wells. NPDES renewal issued 10/28/2009.	O&G
Sunbury Generation	PA0008451	Susquehanna River	3.7	0.08	Snyder	CWT	Taking 80,000 gpd, approved by letter. Application submitted in 10/09 to accept up to 250,000 gpd. Proposing to cease taking wastewater by 12/31/2010. The NPDES amendment application was withdrawn 5/10 and may be resubmitted.	MSW

Facility Name	Permit Number	Receiving Stream	Design Flow (MGD)	Oil & Gas Wastewater Flow (MGD)	County	TYPE[a]	Status / NOTES	Facility Designation[b]
Somerset Regional Water Resources, LLC (SRWR)	PA0253987	East Branch Coxes Creek	1.0	1.0	Somerset	Proposed CWT with 500 mg/L TDS limitation in the permit	The NPDES permit for this facility was issued on December 17, 2009. The Water Quality Management Part II permit application was deficient. The applicant supplied additional information, and the Part II permit is currently under review. SRWR plans to treat for TDS. RO and evaporators are proposed to treat O&G and mine drainage.	PMSW, PO&G
Penn Woods Enterprises, LLC	PA0254037	Casselman River	0.5	0.5	Somerset	Proposed CWT	This application is for issuance of an NPDES permit to discharge treated process water from the Casselman Waterworks Water Treatment Plant in Summit Township, Somerset County.	PO&G, PMSW
Dannic Energy Corp.	PA0233773	Tioga River	0.25	0.25	Tioga	Proposed CWT	Metals precipitation and vacuum evaporator for treatment.	PMSW

Appendix A. (Continued)

Facility Name	Permit Number	Receiving Stream	Design Flow (MGD)	Oil & Gas Wastewater Flow (MGD)	County	TYPE[a]	Status / NOTES	Facility Designation[b]
Big Sandy Oil Company	PA0222011	Allegheny River	0.006	0.004	Venango	CWT	Existing facility which treats only oil well production fluids from own wells.	O&G
Titusville Oil and Gas Associates, Inc. - Hilton Hedley Lease	Approval 6106001 under GP PAG310001	Allegheny River	0.002	0.002	Venango	Proposed CWT	Proposed facility to treat only oil well production fluids from own wells. Approval under general NPDES issued 1/16/2007, no Part II issued and facility not built.	SW, PO&G
PA Brine Treatment - Franklin Facility	PA0101508	Allegheny River	0.205	0.144	Venango	CWT	Renewed permit issued 2/27/2009 authorizing increased discharge rate to 300,000 gpd while maintaining same load limits for chlorides.	O&G, PMSW

Facility Name	Permit Number	Receiving Stream	Design Flow (MGD)	Oil & Gas Wastewater Flow (MGD)	County	TYPEa	Status / NOTES	Facility Designationb
							WQM Permit amendment to reflect modifications necessary to increase discharge rate to 300,000 gpd issued 10/23/2009. The 10/09 Amendment No. 1 to Water Quality Management	
							Permit No. 6182201-T3 authorized the modification and operation of the O&G wastewater treatment facility.	
PA Brine Treatment - Rouseville Facility	PA0263516	Oil Creek	0.08	0.08	Venango	Proposed CWT	Second draft of permit published in PA Bulletin 3/13/2010. EPA withdrew general objection 3/17/2010 based on redraft. Water quality protection report (WQPR) currently being redrafted to reflect proposed TDS regulation changes.	PO&G

Appendix A. (Continued)

Facility Name	Permit Number	Receiving Stream	Design Flow (MGD)	Oil & Gas Wastewater Flow (MGD)	County	TYPE[a]	Status / NOTES	Facility Designation[b]
ARMAC Resources	Approval 6295001 under GP PAG310001	Brokenstraw Creek	0.001	0.001	Warren	CWT	Existing facility which treats only oil well production fluids from own wells.	SW, O&G
Waste Treatment Corporation	PA0102784	Allegheny River	0.213	0.199	Warren	CWT	Renewal application received 10/14/08. Request to increase discharge rate to 400,000 gpd withdrawn. Draft permit published for 30 day comment 1/30/2010.	O&G, PMSW
Mon Valley Brine	PA0253782	Monongahela River	0.2	0.2	Washington	Proposed CWT	O&G wastewaters-CWT	PMSW, PO&G
Borough of California	PA0022241	Monongahela River	1.0	0.0	Washington	POTW	O&G waste flow limited to 1% of Avg Daily flow at POTW by DEP order. This facility no longer accepts large volumes of O&G wastewater.	O&G, MSW

Facility Name	Permit Number	Receiving Stream	Design Flow (MGD)	Oil & Gas Wastewater Flow (MGD)	County	TYPE[a]	Status / NOTES	Facility Designation[b]
Authority of the Borough of Charleroi	PA0026891	Monongahela River	1.6	0.0	Washington	POTW	O&G waste flow limited to 1% of Avg Daily flow at POTW by DEP order. This facility no longer accepts large volumes of oil and gas wastewater.	O&G, MSW
The Washington-East Washington Joint Authority	PA0026212	Chartiers Creek	9.77	0.0977	Washington	POTW	Limiting O&G waste flow to 1% of Avg Daily flow.	O&G, MSW
Reserved Environmental Services, LLC (RES)	PA0254185	Sewickley Creek and Belson Run	Phase I – No Discharge. All water will be recycled to well site. Phase II -1.0 MGD	1.0	Westmoreland	CWT	CWT using municipal drinking water as dilution with recycle back to well sites. RES has a contract with the Westmoreland Water Authority to purchase up to three million gallons of potable water per day to be used as "make-up" water to adjust the concentration of chlorides in the recycled water.	MSW, O&G

Appendix A. (Continued)

Facility Name	Permit Number	Receiving Stream	Design Flow (MGD)	Oil & Gas Wastewater Flow (MGD)	County	TYPE[a]	Status / NOTES	Facility Designation[b]
							RES intends, in Phase I, to treat drilling fluids at the facility and produce an "engineered" controlled fluid/water for recycle to the drilling sites.	
CB Energy	PA0252646	Conemaugh River	0.2	0.2	Westmoreland	CWT	n/a	CBM
CNX Gas Co LLC	PA0252867	Youghiogheny River	0.08	0.08	Westmoreland	CWT	n/a	CBM
CNX Gas Co LLC	PA0253049	Crawford Run / Youghiogheny River	0.08	0.08	Westmoreland	CWT	n/a	CBM
CB Energy Inc	PA0219312	Sewickley Creek	0.2	0.2	Westmoreland	CWT	n/a	CBM
Belden and Blake	PA0218898	Conemaugh River	0.6	0.6	Westmoreland	CWT	n/a	CBM
CB Energy Inc	PA0219452	Crabtree Creek	0.2	0.2	Westmoreland	CWT	n/a	CBM

Facility Name	Permit Number	Receiving Stream	Design Flow (MGD)	Oil & Gas Wastewater Flow (MGD)	County	TYPE[a]	Status / NOTES	Facility Designation[b]
Belden and Blake	PA0218073	Jacobs Creek	0.6	0.6	Westmoreland	CWT	n/a	CBM
Kiski Valley WPCA/McCutcheon Enterprises (CWT)	PA0027626	Kiskiminetas River	7	0.09	Westmoreland	POTW receiving discharge from CWT	Existing CWT facility (McCutcheon Enterprises, which is treating O&G wastewater for discharge to the Kiski Valley WPCA.	PO&G, PMSW
Reserved Environmental Services LLC	WMGR121 SW001	Sewickley Creek	1.0	0.25	Westmoreland	RW Processing	Batch physical & chemical processing – treated wastewater returned to well sites for reuse.	MSW
North Branch Processing, LLC (NBP)	PA0065269	Susquehanna River	0.5	0.50	Wyoming	Proposed CWT	NPDES draft permit mailed 8/4/09. Public hearing held 10/6/09. Working on comment response document. NBP anticipates redrafting and republishing the NPDES permit upon completion of the comment response document.	PMSW

Appendix A. (Continued)

Facility Name	Permit Number	Receiving Stream	Design Flow (MGD)	Oil & Gas Wastewater Flow (MGD)	County	TYPE[a]	Status / NOTES	Facility Designation[b]
Wyoming-Somerset Regional Water Resources LLC (W-SRWR)	PA0065293	Meshoppen Creek at the SR 29 Bridge	0.38	0.38	Wyoming	Proposed CWT	NPDES draft permit mailed 8/4/09. Public hearing held 10/20/09. Working on comment response document. W-SRWR anticipates redrafting and republishing the NPDES permit upon completion of the comment response document.	PMSW

[a] POTW – publically-owned treatment works; CWT – commercial wastewater treatment facility; RW – residual waste.

[b] MSW – Currently accepts Marcellus Shale wastewater; O&G – Currently accepts oil and gas wastewater from formations other than Marcellus Shale; P – proposed facility; CBM – coal bed methane wastewater; SW – water from stripper wells.

Source: Based on PADEP spreadsheet (Furlan 2010).

End Notes

[1] The draft strategy can be downloaded at http://files.dep.statertnership/high_tds_wastewater_strategy_041109.pdf; accessed April 29, 2010.

[2] The May 17 revisions to the rule can be downloaded at http://files.dep.state 0TDS%20Final%20 Rulemaking%20to%20WRAC.pdf; accessed May 20, 2010.

[3] This information was reported in a May 19 article in the *Charleston Gazette*; http://wvgazette.com/News/201005190834; accessed May 21, 2010.

[4] http://aopclearwater.com/The_Clearwater_Process.html; accessed May 3, 2010.

[5] http://files.dep.state.pa.us/Waste/Bureau%20of%20Waste%20Management/WasteMgtPortal Files/SolidWaste/Residual_Waste/GP/WMGR121.pdf; accessed July 3, 2010.

In: Marcellus Shale and Shale Gas
Editor: Gabriel L. Navarro
ISBN: 978-1-61470-173-6

Chapter 3

NATURAL GAS DRILLING IN THE MARCELLUS SHALE NPDES PROGRAM FREQUENTLY ASKED QUESTIONS*

1) What is the Marcellus Shale?

The Marcellus Shale is an organic rich rock that has been estimated to contain from 50 to 500 trillion cubic feet of natural gas[1]. It was deposited in the Appalachian Basin 350 million years ago as part of an ancient river delta and consists of the bottom layer of an Upper Devonian age sedimentary rock sequence. Like most shale, the Marcellus was deposited as extremely fine grained sediment, with small pore spaces and low permeability that prevents gas from easily migrating[1]. Often called the Marcellus Black Shale due to its color, the formation exists under much of southern New York, Pennsylvania, West Virginia, eastern Ohio, and far western Maryland. Although the shale outcrops at its namesake, Marcellus, New York, it generally lies at depths of 5,000 to 9,000 feet throughout much of the area.[2] The Marcellus Shale generally ranges in thickness from 50 to 200 feet.

---

* This is an edited, reformatted and augmented version of a EPA's Office of Wastewater Management publication, dated March 16, 2011.

2) Why is the Marcellus Shale gas extraction suddenly important for natural gas production?

The combination of advances in drilling and fracturing technology, the large volume of natural gas reserves, and its proximity to eastern cities have made the Marcellus Shale an important resource. Although the first commercial shale gas well was drilled in New York in 1821, extensive drilling and extraction of natural gas from shale deposits in the United States did not begin until the 1980's.[3,4] Horizontal drilling techniques, that make gas extraction viable in the Marcellus Shale, did not become commercially available until the late 1980s.[5] Fracturing techniques that are needed to economically extract gas from impermeable shale deposits, like the Marcellus, also recently became refined.[6] Analysis of the Marcellus formation geology suggests that areas in the north central and northeastern regions of Pennsylvania have a high potential to produce significant amounts of gas. This area of the country has not traditionally seen extensive gas well drilling.[7]

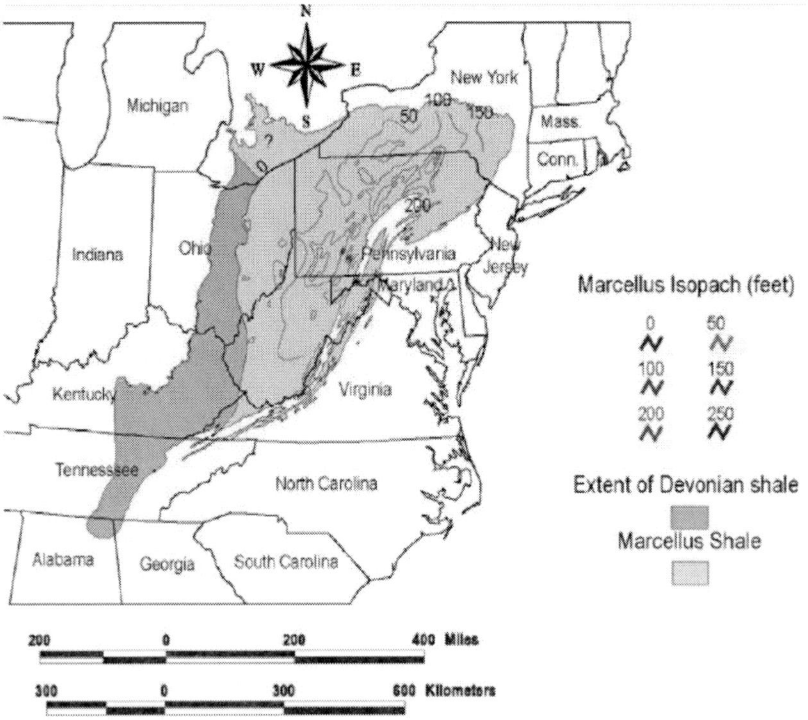

Figure 1. Location of Marcellus Shale[8].

3) How is extraction from the Marcellus Shale different from other natural gas extraction?

Marcellus gas extraction is considered "unconventional" by the Department of Energy's Energy Information Administration because the gas is found within a shale formation rather than a more normal sandstone or limestone rock layer.[9] Conventional gas extraction typically involves drilling through an impervious rock formation into a porous formation saturated with gas and trapped by the impervious cap rock. Conventional extraction typically relies on the high permeability of the rock that allows gas to readily flow to the well for production. Although horizontal wells have become more common over time for conventional gas extraction, wells are more typically relatively straight and vertical.

Unconventional gas extraction includes: deep gas (greater than 15,000 feet), tight gas, shale gas, coal bed methane, gas from geopressurized zones, and methane hydrates. Like tight gas which is extracted from sandstone and limestone deposits that have a low permeability, shale gas extraction requires techniques such as fracturing and horizontal drilling that are less commonly used in conventional extraction. Horizontal drilling is commonly used in shale gas extraction as a means to increase potential production. Horizontal drilling results in a well extending through a much larger portion of the shale; thereby increasing the area from which a well can produce and the amount of gas produced.

In addition to greater use of horizontal drilling, operators make extensive use of hydraulic fracturing as a means to economically produce gas from deposits with low permeability, such as the Marcellus Shale. Hydraulic fracturing requires drillers to pump large amounts of water mixed with sand or other proppants into the shale formation under high pressure (approximately 10,000 psi) to fracture the shale formation adjacent to the wellbore and to create paths that connect the gas to the well. This allows the natural gas to flow freely up the well for compression, transmission, and sale. Once the hydraulic fracturing process is completed and the wellbore pressure is released, approximately one-third of the water flows out of the well[11]. That hydraulic fracturing flowback water (HFFW) must be treated to remove chemicals and minerals.[1] Horizontal wells in the Marcellus Shale require 3 to 5 million gallons for hydraulic fracturing, whereas conventional wells of a similar depth required approximately 1 million gallons of water.[10] The greater quantity of water used for fracturing in shale gas wells is due in part to the extended reach of horizontal wells in addition to the amount of fracturing required to extract gas from a rock that has low permeability[11].

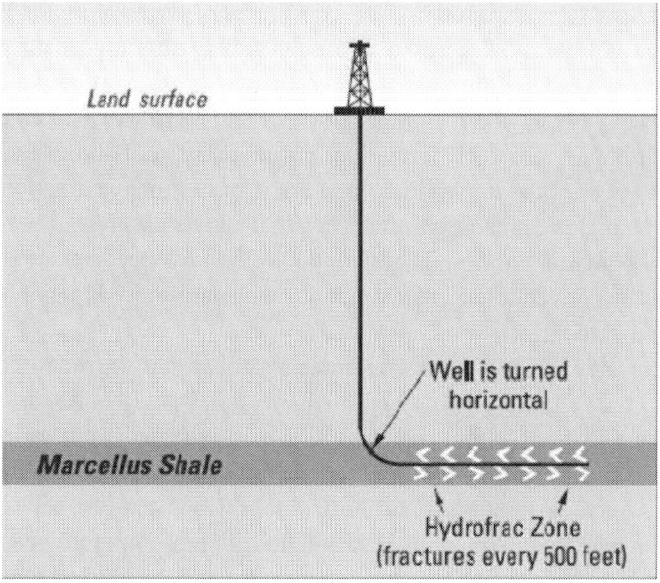

Figure 2. Example of a Horizontal Well[1].

4) How many wells could be expected at a Marcellus gas extraction site?

The number of wells drilled at a site is highly variable and is dependent on local drilling activity, recycling practices of operators, state regulations on well spacing, and local ordinances, among other factors. In general, 1 to 8 wells can be placed on a well pad. A site is expected to consist of only one well pad. Since each well will require numerous trucks to haul away HFFW, a treatment facility (Publicly Owned Treatment Works (POTW) or Centralized Waste Treatment facility (CWT)) would be expected to receive a number of truck loads from a single site.

5) How similar is the Marcellus Shale to other shale deposits where natural gas is currently extracted?

Major shale deposits currently being developed in the United States include the Antrim, Barnett, Fayetteville, Haynesville, Marcellus, and

Woodford Shale. Those shale deposits all have the common characteristic of low porosity and permeability. Extraction almost universally requires horizontal drilling combined with extensive hydraulic fracturing. There are some differences in depth, aerial extent, gas content, and thickness that distinguish between the different shale deposits. A comparison follows in Table 1. Gas extraction activities at all of those shale deposits will present the same challenges for waste disposal.

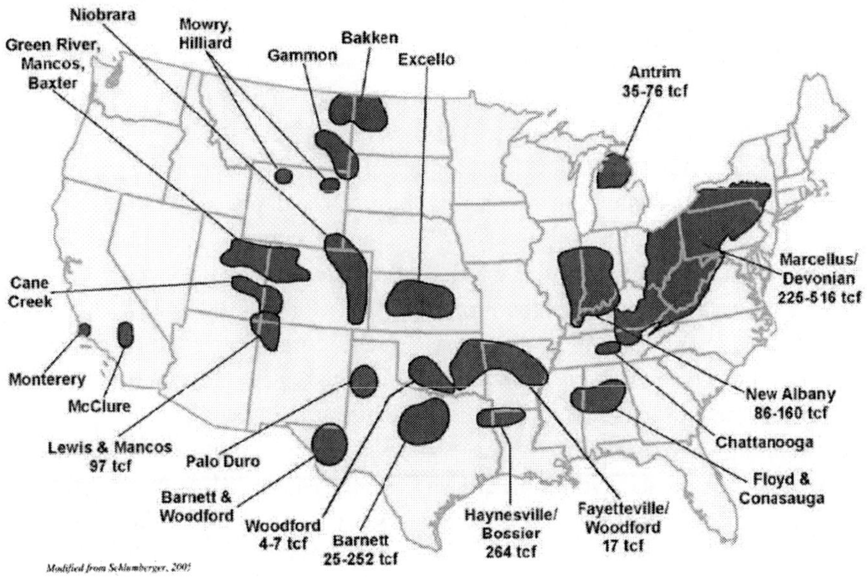

Figure 3. Shale Gas Plays in the United States[12].

Table 1. Comparison of Data for the Gas Shales in the United States[12]

Gas Shale Basin	Estimated Basin Area (mi^2)	Depth (ft)	Net Thickness (ft)	Gas Content (scf/ton)
Antrim	12,000	600-2,200	70-12	40-100
Barnett	5,000	6,500-8,500	100-600	300-350
Fayetteville	9,000	1,000-7,000	20-200	60-220
Haynesville	9,000	10,500-13,500	200-300	100-330
Marcellus	95,000	4,000-8,500	50-200	60-100
Woodford	11,000	6,000-11,000	120-220	200-300

6) Does the Clean Water Act apply to discharges from Marcellus Shale Drilling operations?

Yes. Natural gas drilling can result in discharges to surface waters. The discharge of this water is subject to requirements under the Clean Water Act (CWA).

The CWA prohibits the discharge of pollutants by point sources into waters of the United States, except in compliance with certain provisions of the CWA, including section 402. 33 U.S.C. 1311(a). Section 402 of the CWA establishes the National Pollutant Discharge Elimination System ("NPDES") program, under which EPA, or an authorized state agency, may issue a permit allowing the discharge of pollutants into waters of the U.S.

33 U.S.C. 1342(a). When developing effluent limitations for an NPDES permit, a permit writer must consider limits based on both the technology available to control the pollutants (i.e., technology-based effluent limits) and limits that are protective of the water quality standards of the receiving water (i.e., water quality-based effluent limits). CWA section 301, 33 U.S.C. § 1311; 40 CFR 125.3(a). The technology-based requirements for direct discharges from oil and gas extraction facilities into surface waters are found in 40 Code of Federal Regulations (CFR) Part 435 (see question7, below).

In addition to direct discharges, wastewaters may be indirectly discharged into waters of the U.S. through sewer systems connected to publicly owned treatment works (POTW) that discharge directly to waters of the U.S. or by being introduced by truck or rail into a POTW that discharges directly. EPA regulations set standards for the pretreatment of wastewater introduced to a POTW including prohibiting introduction of wastes that interferes with, passes through or are otherwise incompatible with POTW operations. 33 U.S.C. § 1317(b)(1). EPA has developed other nationally applicable pretreatment standards under section 307(b) in its General Pretreatment Regulations for Existing and New Sources of Pollution (Pretreatment Regulations) at 40 C.F.R. Part 403. These pretreatment standards are applicable to any user of a POTW, defined as a source of an indirect discharge. 40 C.F.R. 403.3(h). These national pretreatment standards include: 1) a general prohibition and 2) specific prohibitions. 40 C.F.R. 403.5.(a)(1) and (b). The general prohibition prohibits any user of a POTW to introduce a pollutant into the POTW that will cause pass through or interference. The regulations define both pass through and interference. Section 307(d) of the Act prohibits discharge in violation of any pretreatment standard. 33 U.S.C. § 1317(d). See questions 10 and11, below, for additional information on pretreatment requirements.

Wastewater may also be disposed of at centralized waste treatment facilities (CWTs). Technology-based standards for CWTs can be found at 40 CFR Part 437. Issues and requirements associated with CWTs are discussed below under questions13, 14 and 15.

7) Do the Oil and Gas Extraction effluent guidelines for onshore operations, found at 40 CFR Part 435, Subpart C, apply to Marcellus Shale gas drilling?

Yes. The technology-based regulations (40 CFR Part 435, Subpart C) apply to onshore facilities "engaged in the production, field exploration, drilling, well completion and well treatment in the oil and gas extraction industry." Gas drilling in the Marcellus Shale fits squarely within this applicability statement. Although, as discussed in Question 3 above, Marcellus Shale gas extraction may be considered "unconventional" gas extraction, the wastestreams generated by processes used in such extraction, such as hydraulic fracturing, were considered and covered by the effluent guideline. *See, e.g.* 41 Fed. Reg. 44946 (Oct. 13, 1976); Technical Development Document at 22-23, 96, 137. Accordingly, the discharge prohibitions in 40 CFR Part 435, Subpart C, apply to Marcellus Shale gas extraction.

The effluent guidelines at 40 CFR 435, Subpart C establish best practicable control technology currently available (BPT) requirements for onshore facilities: "there shall be no discharge of waste water pollutants into navigable waters from any source associated with production, field exploration, drilling, well completion or well treatment (i.e., produced water, drilling muds, drill cuttings, and produced sand)." During the issuance process for the guidelines, EPA identified different technologies that operators can use to comply with this technology-based regulation (e.g., underground injection, use of pits/ponds for evaporation).

8) Since 40 CFR Part 435, Subpart C applies to the Marcellus Shale drilling activity, may an NPDES permit authorize onsite discharge of this wastewater to a water of the U.S.?

No. Because all applicable technology based requirements must be applied in NPDES permits under the CWA section 402(a) and implementing regulations at 40 CFR 125.3, an NPDES permit issued for the drilling activity would need to be consistent with 40 CFR Part 435, Subpart C, which states that 'there shall be no discharge of wastewater pollutants into navigable waters from any source associated with production, field exploration, drilling, well completion, or well treatment (i.e., produced water, drilling muds, drill cuttings, and produced sand)." 13

9) Are facilities subject to 40 CFR Part 435, Subpart C required to obtain an NPDES permit that imposes the "no discharge" requirement for the activities identified in Subpart C?

No. EPA's regulations at 40 CFR 122.21(a) require permits only for facilities that "discharge or propose to discharge." Accordingly, facilities subject to a "no discharge"

limit that do not discharge or propose to discharge are not required to apply for NPDES permits. States can use their own authority to ensure that the no discharge requirement in the effluent guideline is properly applied and to ensure that operator compliance is demonstrated.

Facilities subject to a zero discharge requirement may apply for permit coverage to qualify for the upset or bypass defense in the event of an unanticipated discharge resulting from an exceptional incident that otherwise would trigger a CWA Section 301 violation for discharging without a permit. *See* 40 CFR 122.41(m) and (n).

10) May Shale Gas extraction (SGE)[14] wastewaters be discharged to a POTW?

POTWs may accept SGE wastewater under certain circumstances. Process wastewater from such operations may be introduced to POTWs but only to the extent that such wastewater discharges are in compliance with all Federal, State, and local requirements governing the introduction of such wastewaters into the POTW. EPA has generally promulgated pretreatment standards that apply to wastewater introduced to POTWs along with effluent guideline for industrial categories.

The current Federal regulations at 40 CFR 435, Subpart C do not include pretreatment standards that address the disposal of Marcellus Shale wastewater to POTWs. However, EPA's General Pretreatment regulations prohibit the introduction of wastewater into a POTW in certain defined circumstances, including the introduction of any pollutants which "pass through" or cause "interference" with POTW operations. 40 CFR Part 403.3(k)(1) defines interference as inhibiting or disrupting the POTW, its treatment processes or operations, or its sludge processes, use or disposal. Therefore, in addition to prohibiting the introduction of pollutants into the POTW that would disrupt the treatment process, the general regulations also prohibit the introduction of pollutants in concentrations that contaminate biosolids and make them inconsistent with the POTW's chosen method of use or disposal. Pass through is defined at 40 CFR 403.3(p) to mean "a discharge which exits the POTW into waters of the United States in quantities or concentrations which, alone or in conjunction with a discharge or discharges from other sources, is a cause of a violation of any requirement of the POTW's NPDES permit (including an increase in the magnitude of a violation)." All non-domestic discharges must comply with these requirements. See 40 CFR 403.5(a) and (b).

Note: SGE wastewater that is discharged to a POTW from a CWT may have the same issues as wastewater taken directly to a POTW from a shale gas extraction well and pass through and interference will also need to be addressed.

11) What requirements do POTWs need to meet in order to accept shale gas wastewater?

POTWs need to comply with their NPDES permit terms and conditions. In accordance with the NPDES permitting regulation at 40 CFR 122.42(b)[15], permits must include conditions that require - - -"all POTWs must provide adequate notice to the Director [EPA and/or the state NPDES permitting/pretreatment authority[16]] of the following:

1) Any new introduction of pollutants into the POTW from an indirect discharger which would be subject to section 301 or 306 of the CWA if it were directly discharging those pollutants; and
2) Any substantial change in the volume or character of pollutants being introduced into that POTW by a source introducing pollutants into the POTW at the time of issuance of the [POTW's] permit.
3) For the purposes of this paragraph, adequate notice shall include information on (i) the quality and quantity of effluent introduced into the POTW, and (ii) any anticipated impact of the change on the quantity or quality of effluent to be discharged from the POTW."

To the extent that a permit so requires, when considering the acceptance of such wastewater, a POTW needs to collect information from the industry on the quality and quantity of the SGE wastewater proposed to be introduced to the POTW and assess the potential impact to the POTW if the POTW were to accept the wastewater. For SGE wastewater, that discharge characterization should include the concentrations of total dissolved solids, specific ions, such as chlorides and sulfate, specific radionuclides, metals, and other pollutants that could reasonably be expected to be present in wastewater from a well. In addition to the ions, radionuclides, and metals that can be expected to be present in wastewater produced from a well, the characterization should include all chemicals used in well drilling, completions, treatment, workover, or production, that could reasonably be expected to be present in wastewater. Pursuant to the permit, this information must generally be reported to EPA and/or the State program before the POTW may accept the HFFW. "Adequate notice" is meant to provide the EPA (or the state NPDES permitting authority) with enough time to determine if the POTW NPDES permit needs to be modified in order to address potential effects due to the potential new indirect discharger. In cases such as Pennsylvania, where the state is the permitting authority and EPA is the approval authority for pretreatment, the POTW must submit the required information to both agencies. In addition to this notification, all industrial user discharges to a POTW must comply with the specific prohibitions of 40 CFR 403.5(b), any applicable categorical standards, and any state and local limits.

EPA Regions, in their oversight role, should work with authorized States to ensure that NPDES permits for POTWs include the pretreatment notification requirements and definitions of 40 CFR 122.2, 122.42(b), and 403.5(b). By including those requirements in permits, the permitting authorities will help prevent potential oversights of the notification requirements by POTW operators.

EPA recognizes that POTW operations vary due to site-specific factors. All POTWs with approved pretreatment programs, and all other POTWs designated by EPA or the state as having experienced or having the potential to experience pass through or interference, must develop technically-based local limits where necessary to comply with the general pretreatment standards. See 40 CFR 403.5(c) & 403.8(a). To assist in this evaluation, EPA has issued guidance on establishing local limits: Local Limits Development Guidance, EPA-833-R-04-002A, July 2004.17 18

12) What are the main potential pollutants of concern for POTWs accepting SGE wastewaters?

Constituents in SGE wastewater such as total dissolved solids (TDS) have been found to be present at concentrations ranging from 280 mg/l to 345,000 mg/l.[19] Chloride has been reported in concentrations up to 196,000 mg/l.[20] TDS is not significantly removed by most conventional POTW treatment systems; therefore, pretreatment of the wastewater would be required prior to discharge to the POTW. However, very little comprehensive data have been collected nationwide on TDS treatment capability at POTWs. Common constituents of TDS include calcium and magnesium (also a measure of "hardness"), phosphates, nitrates, sodium, potassium, sulfates, chloride, and even barium, cadmium, and copper. A literature data search revealed that some of these individual constituents of TDS may result in POTW process inhibition in activated sludge, nitrification, and anaerobic digestion processes. POTWs may exhibit these process inhibitions from these individual constituents at concentrations that are several magnitudes lower than the composite TDS found in SGE wastewater (example: sulfate at 400-1000 mg/l disrupting anaerobic digestion processes; chloride at 180 mg/l disrupting nitrification processes[21]). High concentrations of chlorides, such as in Marcellus SGE wastewater, can disrupt biological treatment units. Some POTWs that had previously accepted oil and gas extraction waste through their pretreatment programs experienced operational problems due to high concentrations and spikes in concentrations of TDS.[22] In addition, some of the constituents in oil and gas extraction waste, such as metals, can precipitate during the treatment process and contaminate biosolids which may require expensive decontamination of biosolids drying beds or change the chosen method of use or disposal. Bromide, which can be present in SGE wastewater in significant concentrations, has the potential to be present in POTW effluent as a disinfection byproduct and may cause an increase in whole effluent toxicity[21].

Because there is a significant possibility that SGE wastewater may "pass through" the POTW, causing the POTW to violate its permit, cause "interference" with the POTW's operation, or contamination of biosolids, acceptance of the waste is not advisable unless it's effects on the treatment system are well understood and the wastewater is not reasonably expected to cause pass through or interference. POTWs cannot accept Marcellus wastewater if acceptance of the wastewater would result in violations of the POTW's permit, the POTW's requirement under 40 CFR 403.5(c) to develop and enforce local limits to implement the general and specific prohibitions of 403.5(a)(1) and (b), or contamination that interferes or disrupts biosolids processes, uses, or disposal. NPDES permits for discharges from POTWs to water of the U.S. also must meet applicable water quality-based requirements that are discussed in more detail in question number 21.

Radionuclides in Marcellus SGE wastewater also pose a challenge for POTWs. Radionuclides are discussed below in the response to question number 19.

These same pollutants may be of concern to POTWs that accept wastewater from CWTs that themselves accept SGE wastewaters. Many CWTs may not effectively treat SGE wastewater. Appropriate limits and pretreatment requirements will need to be developed by the permitting authority and the pretreatment control authority.

13) Could SGE wastewater be transferred to a CWT facility for treatment and discharge?

Yes. Although the direct discharge of wastewater from drilling operations is not authorized, the wastewater may be transported to a CWTs for treatment and subsequent discharge. Discharges from a CWT are subject to the effluent limitations guidelines and standards established under 40 CFR Part 437.

Additional limits may be required to address pollutants in the wastewater that were not considered in developing the CWT effluent guideline. For such pollutants, EPA's NPDES regulations require that permit writers include technology-based limits developed on a case-by-case, "best professional judgment" (BPJ) basis. *See* 40 CFR §125.3(c)(3) ("Where promulgated effluent limitations guidelines only apply to certain aspects of the discharger's operation, or to *certain pollutants*, other aspects or activities are subject to regulation on a case by case basis..."). In developing technology-based BPJ limits, the permit writer must consider the factors specified in 40 CFR 125.3(d), the same factors that EPA considers in establishing categorical effluent guidelines.

In developing the CWT effluent guideline, EPA did not evaluate certain pollutants that are likely to be present in SGE wastewater, such as radionuclides. Consequently, the permitting authority will need to develop best professional judgment technology based effluent limits to address those pollutants identified in the effluent but not considered by the CWT Effluent Guidelines and incorporate these limits in the CWT's NPDES permit.

For some pollutants, such as total dissolved solids (TDS), EPA considered, but did not establish, pollutant limitations in the effluent guidelines. TDS levels in Marcellus Shale wastewaters have been measured to be present in concentrations up to 345,000 mg/l[20]. High concentrations of TDS will require advanced waste water treatment, such as distillation, and may cause scaling which requires frequent cleaning of equipment[10]. In addition to any applicable technology-based requirements, NPDES permits for discharges from CWTs to waters of the U.S. also must meet applicable water quality-based permitting requirements. See question number 21 for more detail on water quality permitting.

14) What Subpart of 40 CFR Part 437 should be used for the Marcellus Shale wastewater?

40 CFR Part 437 includes three subparts to address different industries that may dispose of wastewater in a CWT. Those subparts include: Metals Treatment and Recovery, Oils Treatment and Recovery, and Organics Treatment and Recovery. When the Effluent Limitations Guidelines were promulgated, EPA understood that industrial wastes would not always clearly fit under one of the subcategories. To address the issue of categorization of wastewater, EPA developed guidance for permit writers to determine which subpart of the 40 CFR Part 437 ELGs best addresses waste accepted by a CWT.[23] Chapter 5 of the guidance lists different waste sources that were examined during development of the ELG and were determined to best be addressed under each subpart. For waste sources not listed, the guidance contains additional criteria based on oil and grease content and metals concentrations that can be used for this determination. Available data for Marcellus shale extraction waste water show that the waste does not fit under the Oils or Metals Subcategories. The guidance suggests regulating waste under the Organics Subcategory for cases where it does not fit under the other Subcategories.[23] However, this determination was made only using Marcellus shale waste data. CWTs are expected to receive waste containing different pollutant types and concentrations originating from a variety of sources.[24] The permit writer will need to reexamine this determination based on site specific information when drafting a permit.

15) How is transportation of waste by pipeline addressed by the CWT regulations?

CWTs may accept wastewater transported to the CWT via pipeline. The CWT would be subject to applicable limitations imposed on its discharge through its NPDES permit or pretreatment program control mechanism. The CWT ELGs are only applicable to CWT discharges of treated piped wastewater if the treated piped wastewater is comingled with other wastewater covered by the CWT ELG. If the piped wastes are not commingled, the permitting authority will need to develop best professional judgment technology based effluent limits for discharges of piped wastewater from the CWT. The CWT regulations at 40 CFR 437.1(b)(3) address waste received via pipeline from offsite as follows:

> "(b) This part does not apply to the following discharges of wastewater from a CWT facility:
>
> . . . (3) Wastewater from the treatment of wastes received from off-site via conduit (e.g., pipelines, channels, ditches, trenches, etc.) from the facility that generates the wastes unless the resulting wastewaters are commingled with other wastewaters subject to this provision. A facility that acts as a waste collection or consolidation center is not a facility that generates wastes."

The requirement was included in the regulations to address wastes that are not as variable as those that were typically found to be treated at the CWT facilities studied during development of the ELGs. Unlike traditional CWT facilities, pipeline customers and wastewater sources do not change and are limited by the physical and monetary constraints associated with pipelines. In addressing this issue, the preamble to the proposed regulation states:

> "EPA has concluded that the effluent limitations and pretreatment standards for centralized waste treatment facilities should not apply to such pipeline treatment facilities because their wastes differ fundamentally from those received at centralized waste treatment facilities. In large part, the waste streams received at centralized waste treatment facilities are more concentrated and variable, including sludges, tank bottoms, off-spec products, and process residuals. The limitations and standards developed for centralized waste treatment facilities, in turn, reflect the types of waste streams being treated and are necessarily different from those promulgated for discharges resulting from the treatment of process wastewater for categorical industries."[25]

This issue was also addressed in the final rule which further clarified that waste delivered via pipeline would have a more uniform flow rate and with a relatively consistent pollutant concentration. Wastes delivered solely by pipeline would be more consistent with a traditional manufacturing facility that did not accept waste from a variety of different sources.[26]

16) What potential hazardous waste issues apply to the acceptance of oil & gas extraction wastewater at a POTW or CWT via truck, train, or dedicated pipe?

Waste generated by activities associated with the exploration, development, and production of crude oil or natural gas, at primary field operations, are exempt from regulation under RCRA Subtitle C. See 40 CFR 261.4(b)(5). See also the July 1988 Regulatory Determination (53 FR 25466) and the March 1993 clarification of the Regulatory Determination (58 FR 15284) at http://www.epa.gov/epawaste/nonhaz/ industrial/special/oil/ index.htm. These wastes include drilling fluids, produced water, and other wastes associated with the exploration, development, or production of crude oil or natural gas. According to the legislative history, the term "other wastes associated" specifically includes waste materials intrinsically derived from primary field operations associated with the exploration, development, or production of crude oil and natural gas (e.g., spent hydraulic fracturing fluids). The exemption does not apply to excess supplies, such as unused drilling fluids or treatment chemicals. POTWs or CWTs receiving exempt oil and gas extraction wastewaters would not be receiving hazardous wastes and thus would not need to meet RCRA hazardous waste requirements, including RCRA permit or permit-by-rule requirements. For additional clarity on this issue regarding the status of oil and gas exploration and production wastes that are exempt from RCRA subtitle C regulations, see: http://www.epa.gov/epawaste/nonhaz/industrial/special/oil/oil-gas.pdf.

17) Does Part 435 Subpart G apply to the treatment and discharge of wastewaters from the Onshore Subcategory if those wastewaters were sent off-site for treatment and discharge at a facility covered by another ELG, such as a Centralized Waste Treatment (CWT) facility under Part 437?

No. EPA promulgated Subpart G, in part, to eliminate the practice of sending wastewaters from one Part 435 subcategory to another to take advantage of less stringent discharge requirements. Thus, for example, a facility regulated by the Coastal subcategory limitations located near a facility subject to the Offshore subcategory limitations might have sent its wastewater for treatment at the Offshore facility in order to get around the no discharge requirements. Under Subpart G, even if the Coastal subcategory facility transports its wastewater for treatment and/or disposal at the Offshore subcategory facility, the discharge would still be subject to the more stringent no discharge limitations for discharge to navigable waters.

If, however, an Onshore subcategory facility transports its wastewaters to an off-site centralized waste treatment facility, Subpart G would not apply. In this case, the wastewater discharge would be regulated by Part 437. *See* 40 CFR §437.1 (providing that Part 437 applies to "[t]reatment and recovery of ... industrial metal-bearing waters, oily wastes and organic-bearing wastes received from off-site"). In this scenario, transferring wastewaters off-site for authorized disposal meets the no discharge requirement in Part 435 Subpart C ("no discharge of waste water pollutants into navigable waters").

18) What is the definition of "off-site" in regard to SGE wastewater treated at CWTs?

From 40 CFR 122.2: *Site* means the land or water area where any "facility or activity" is physically located or conducted, including adjacent land used in connection with the facility or activity. *Facility or activity* means any NPDES "point source" or any other facility or activity (including land or appurtenances thereto) that is subject to regulation under the NPDES program.

For gas drilling activities, the land identified in the drilling permit; including the locations of wells, access roads, lease areas, and any lands where the facility is conducting its exploratory, development or production activities, or adjacent lands used in connection with the facility or activity, would constitute the site. Land that is outside the boundaries of that area is considered to be "off-site." (see also 40 CFR 437.2(n)

19) The Marcellus Shale is often referred to as a radioactive black shale in literature[27]. Are radionuclides an issue of concern with natural gas extraction and wastewater disposal?

Radionuclides associated with oil and gas extraction, also referred to as Naturally Occurring Radioactive Material (NORM), are a long standing waste management issue. Many states have addressed the issues associated with NORM in oil and gas extraction through their regulatory programs.[28,30,6] Radionuclides often exist in low concentrations in oil and gas waste and have been found to form deposits over time in piping and equipment. The issues commonly related to radionuclides in oil and gas extraction waste are decontamination of equipment and human health risk for workers.[29,30] Several states with extensive oil and gas extraction activity have also developed requirements for disposal facilities that accept radionuclide contaminated waste.[28] Since oil and gas extraction waste is not discharged in many states, water quality and human health issues associated with discharges under NPDES permits have not been been extensively examined.

The Marcellus Shale has been found to contain NORM that can be in fairly high concentrations in oil and gas extraction wastewater. Radium 226 has been found to be present in concentrations up to 16,030 pCi/l in Marcellus Shale produced water.[31] HFFW from the Marcellus Shale has not been monitored extensively for radionuclides; however, Alpha particles have been found to be present at concentrations up to 18,950 pCi/l.[31] Those radionuclide concentrations exceed the drinking water Maximum Contaminant Levels of 5 pCi/L for Radium 266 and 15 pCi/l for Alpha particles. Although few studies are available that would help to understand the issue of NORM in POTW or CWT effluent, EPA is working with Pennsylvania to gather information and determine whether additional permit limits are needed to protect downstream drinking water supplies.

Based on existing information on NORM associated with oil and gas extraction, it appears that care should be taken to address impacts to treatment facilities, such as scale buildup in equipment and contamination of sludge [biosolids]. Contamination of biosolids at POTWs that requires a change of disposal practice (e.g., radioactivity, etc.) is considered to be interference under the pretreatment program. See 40 CFR 403.3(k)(2) and 403.5(a)(1).

The discharge of shale gas wastewater from POTWs or CWTs has the potential to result in a discharge of radioactive contaminants. Such discharges must be characterized to determine whether reasonable potential exists for impacts to downstream Public Water Systems and other applicable water quality standards. If so, appropriate permit limits must be established.

When the 40 CFR Part 437 effluent limitations guidelines were developed, EPA found that CWTs were not designed to remove radionuclides. Many CWTs also discharge to POTWs rather than directly discharging to Waters of the United States. The same issues that apply to POTWs accepting wastewater from gas well operators also apply to wastewater accepted from CWTs[32].

20) Can any of the Marcellus Shale gas extraction activity fall under Part 435 Subpart F – Stripper Subcategory?

No. The Stripper Subcategory is clearly limited to onshore facilities which produce 10 barrels per well per calendar day or less of crude oil. The Marcellus Shale activity is gas extraction.

21) What water quality-based requirements may apply in NPDES permits for discharges of Marcellus Shale Wastewater from POTWs and CWTs to waters of the U.S.?

EPA's NPDES regulations also require permit writers to include any more stringent requirements necessary to meet applicable water quality standards. Specifically, the regulations require limits to control all discharges that have the reasonable potential to cause or contribute to exceedences of water quality standards. 40 CFR 122.44(d)(1)(i). Accordingly, where, after application of technology-based effluent limits, the discharge of Marcellus Shale wastewater has a reasonable potential to cause or contribute to exceedences of water quality standards, the permit writer will need to develop water quality based effluent limits (WQBELs) for the POTWs or CWT's NPDES permit to protect water quality. Additional requirements may be needed to comply with other State regulations.[33]

WQBELs may be needed for TDS, in particular, where discharges of the pollutant from CWTs or POTWs have the reasonable potential to exceed state numeric or narrative water quality criteria. Since few states have established numeric water quality criteria for TDS, permitting authorities may need to develop a numeric translator to protect the state's narrative water quality criteria. In the Marcellus Shale wastewater, chloride typically constitutes about 50% of the total makeup of Total Dissolved Solids (TDS) in a sample. Elevated chloride levels can interfere with an aquatic organism's ability to maintain osmotic balance/control with its environment, as well as cause other effects. Some states have applicable numeric water quality criteria for chloride. Where a state has a numeric criterion, NPDES permit regulations require that permitting authorities assess reasonable potential and established permit limits where necessary to protect water quality based on the applicable numeric criterion. Where a state has not developed a numeric criterion for chloride, EPA recommends that permitting authorities use a numeric translation of the applicable narrative criterion pursuant to 40 CFR 122.44(d)(1)(vi). In developing such translation, EPA recommends using EPA's current 304(a) national recommended criteria for chloride for protection of aquatic life. These criteria were published by EPA in 1988. The current national criteria for Chloride are: acute aquatic life criteria of 860 mg/l, and chronic aquatic life criteria of 230 mg/L. EPA is currently in the process of updating these recommended criteria to reflect the latest science. That update is expected to be proposed by the end of 2011 and finalized in 2012.

22) Does EPA's storm water definition at 40 CFR 122.26(b)(14)(iii) include discharges from a natural gas drilling operation?

40 CFR 122.26(b)(14)(iii) does include natural gas activities, but only to the extent that they require permit coverage as described in 122.26(a)(2)(ii) and 122.26(c)(1)(iii).

In general, the Director may not require a permit for discharges of storm water from any field activities or operations associated with oil and gas exploration, production, processing, or treatment operations or transmission facilities, including activities necessary to prepare a site for drilling and for the movement and placement of drilling equipment, whether or not such field activities or operations may be considered to be construction activities.[34]

Exceptions to the above general exemption may be found at 122.26(c)(1)(iii), which states: "*The operator of an existing or new discharge composed entirely of storm water from an oil or gas exploration, production, processing, or treatment operation, or transmission facility is not required to submit a permit application in accordance with paragraph (c)(1)(i) of this section, unless the facility:*

a) Has had a discharge of storm water resulting in the discharge of a reportable quantity for which notification is or was required pursuant to 40 CFR 117.21 or 40 CFR 302.6 at anytime since November 16, 1987; or
b) Has had a discharge of storm water resulting in the discharge of a reportable quantity for which notification is or was required pursuant to 40 CFR 110.6 at any time since November 16, 1987; or
c) Contributes to a violation of a water quality standard."

While oil and gas-related construction is subject to the conditional exemption, operators should still implement best management practices when undertaking earth disturbing activities to prevent discharging pollutants, including sediment, that would cause or contribute to water quality violation, and which would trigger storm water permitting requirements.

General Note

These Q&As provide advice on how to issue National Pollutant Discharge Elimination System permits for discharges from natural gas drilling in the

Marcellus Shale. These Q&As do not impose legally binding requirements on EPA, states, tribes, other regulatory authorities, or the regulated community, and may not apply to a particular situation based upon the circumstances. EPA, state, tribal and other decision makers retain the discretion to adopt approaches on a case-by-case basis that differ from those provided in the Q&As where appropriate. EPA may update these Q&As in the future as better information becomes available.

End Notes

[1] Soeder, D.J., and Kappel, W.M., 2009, Water Resources and Natural Gas Production from the Marcellus Shale: U.S. Geological Survey Fact Sheet 2009–3032, 6 p.

[2] USGS, 2006, Assessment of Appalachian Basin Oil and Gas Resources: Devonian Shale-Middle and Upper Paleozoic Total Petroleum System.

[3] Hill, D.G., etal, 2003 , Fractured Shale Gas Potential in New York, posted at: http://www.pe.tamu.edu/wattenbarger/public_html/Selected_papers/--Shale%20Gas/fractured%20shale%20gas%20potential%20in%20new%20york.pdf

[4] Shirley, K., 2001, Shale Gas Exciting Again, Explorer, posted at: http://www.aapg.org/explorer/2001/03mar/gas_shales.cfm

[5] Energy Information Administration, Office of Oil and Gas, U.S. Department of Energy, Drilling Sideways – A Review of Horizontal Well Technology and its Domestic Application, April, 1993.

[6] Ground Water protection Council, 2009, Modern Shale Gas Development in the United States: A Primer, 116 p., posted at: www.gwpc.org

[7] Reference: "Drilling for Natural Gas in the Marcellus Shale Formation - Frequently Asked Questions" as written by the Pennsylvania Department of Environmental Protection and posted at http://www.dep.state forms/marcellus/marcellus.htm.

[8] Milici, R.C., USGS Open File Reports 2005-1268, Assessment of Undiscovered Natural Gas Resources in Devonian Black Shales, Appalachian Basin, Eastern United States, 2005

[9] See http://www.eia.doe.gov/oiaf/analysispaper/unconventional_gas.html.

[10] Gaudlip, A.W., et. al., 2008, Marcellus Shale Water Management Challenges in Pennsylvania, SPE 119898

[11] University of Maryland, Reconciling Shale Gas Development with Environmental Protection, Landowner Rights, and Local Community Needs, Schools of Public Policy, July, 2010.

[12] Arthur, J., et.al., 2008, An Overview of Modern Shale Gas Development in the United States, ALL Consulting, 21 p., posted at: http://www.all-llc.com/publicdownloads/ALLShaleOverviewFINAL.pdf

[13] Note: Shale gas wells from other formations that are located west of the 98th meridian may be regulated under the Agriculture and Wildlife Water Use Subcategory of the Oil and Gas Extraction Category (40 CFR Part 435, Subpart E). Produced water discharges can be authorized under that subcategory if they are of good enough quality to be used by agriculture or wildlife watering and actually are put to that use. The subcategory only allows the discharge of produced water. The discharge of all other waste streams, such as completion fluids, cannot be authorized under Subpart E.

[14] SGE wastewater includes HFFW, produced water, spent drilling fluids, and spent well completion and treatment fluids that have result from shale gas extraction activities.

[15]Applicable to State NPDES programs, see 40 CFR 123.25.

[16] Under 40 CFR 122.2, "*Director*means the Regional Administrator or the State Director, as the context requires, or an authorized representartive. When there is no "approved State program" and there is an EPA administered program, "Director" means the Regional Administrator." Where a State not have an approved State pretreatment program, the Regional Administrator is the Director of the pretreatment program under this provision.

[17]Available at: http://cfpub.epa.gov/npdes/home.cfm?program_id=3

[18] Guidance Manual for the Control of Wastes Hauled to Publicly owned treatment works" EPA 833-B-98- 003, September 1999.

[19] Haynes, Thomas, 2009, Sampling and Analysis of Water Streams Associated with the Development of Marcellus Shale Gas, Gas Technology Institute, Des Plaines, IL.

[20] NYSDEC, 2009, Supplemental Generic Environmental Statement on the Oil, Gas, and Solution Mining Regulatory Program, Well Permit Issuance for Horizontal Drilling and High-Volume Fracturing to Develop the Marcellus Shale and Other Low Permeability Gas Reservoirs, Appendix 13, available at: http://www.dec.ny.gov/energy

[21] USEPA, Local Limits Development Guidance Appendices, EPA 833-R-04-002B, July, 2004

[22] Record of communications between Scott Wilson (EPA, OWM), Morgan City, LA pretreatment program, and Ted Palit (EPA Region 6)

[23] USEPA, Small Entity Compliance Guide, Centralized Waste Treatment Effluent Limitations Guidelines and Pretreatment Standards (40 CFR Part 437), EPA-821-B-01-003, June, 2001, posted online at: http://www.epa.gov/waterscience/guide/cwt/CWTcompliance_guide.pdf

[24] 64 FR 2286, January 13, 1999

[25] 60 FR 5463 - 5506, January 27, 1995

[26] 65 FR 81241 - 81313, December 22, 2000

[27] Milici, R.C. and C.S. Sweeney, 2006, Assessment of Appalachian Basin Oil amd Gas Resources: Devonian Shale – Middle and Upper Paleozoic Total Petroleum System, Open File Report Series 2006- 1237, U.S. Department of Interior, USGS.

[28] http://norm.iogcc.state

[29] USGS, 1999, Naturally Occurring Radioactive Materials (NORM) in Produced Water and Oil Field Equipment – an Issue for the Energy Industry, USGS Fact Sheet FS-142-99

[30] Railroad Commission of Texas, NORM – Naturally Occurring Radioactive Material, posted at: http://www.rrc.state

[31] NYSDEC, 2009, Supplemental Generic Environmental Statement on the Oil, Gas, and Solution Mining Regulatory Program, Well Permit Issuance for Horizontal Drilling and High-Volume Fracturing to Develop the Marcellus Shale and Other Low Permeability Gas Reservoirs, Appendix 13, available at: http://www.dec.ny.gov/energy

[32] *Development Document for the CWT Point Source Category*, Final Rule: Development Document, USEPA, Washington, DC, 2000, available online at:. http://water

[33] PA Environmental Quality Board, Proposed Rulemaking, 39 Pa.B. 64671, November 7, 2009 Available at: http://www.pabulletin.com/secure/data/vol39/39-45/2065.html

[34] See CWA section 402(l)(2) and CWA section 502(24) as amended by the Energy Policy Act of 2005 section 323

In: Marcellus Shale and Shale Gas
Editor: Gabriel L. Navarro

ISBN: 978-1-61470-173-6

Chapter 4

IMPACT OF THE MARCELLUS SHALE GAS PLAY ON CURRENT AND FUTURE CCS ACTIVITIES*

United States Department of Energy

DISCLAIMER

This report was prepared as an account of work sponsored by an agency of the United States Government. Neither the United States Government nor any agency thereof, nor any of their employees, makes any warranty, express or implied, or assumes any legal liability or responsibility for the accuracy, completeness, or usefulness of any information, apparatus, product, or process disclosed, or represents that its use would not infringe privately owned rights. Reference therein to any specific commercial product, process, or service by trade name, trademark, manufacturer, or otherwise does not necessarily constitute or imply its endorsement, recommendation, or favoring by the United States Government or any agency thereof. The view and opinions of authors expressed therein do not necessarily state or reflect those of the United States Government or any agency thereof.

---

* This is an edited, reformatted and augmented version of a National Energy Technology Laboratory publication, from www.netl.doe.gov, dated August 2010.

1.0. INTRODUCTION

The Marcellus Shale is a major geologic formation underlying significant portions of New York, Ohio, Pennsylvania, and West Virginia. Although it is a very tight formation, it contains a massive quantity of natural gas, thus making it of great economic importance. This paper covers the geology of the Marcellus Shale (extent, depth, gas producing potential, properties, etc.), the techniques used to produce the gas, and the potential for Carbon Capture and Storage (CCS) in the Marcellus Shale or adjacent formations. Because of the low permeability of shale units, hydraulic fracturing and horizontal drilling were developed in the Barnett Shale of Texas during the 1990's; these were the key enabling technologies that made recovery of shale gas economically viable. These technologies have been applied to the Marcellus Shale and other shale gas basins. In addition to gas production from the Marcellus Shale and other gas shale basins in the U.S., this paper discusses the impact of shale gas exploration and production on the potential for CCS in the Marcellus and other units in the Appalachian Basin.

Because of its large extent and volume, as well as the possibility of using some of the existing infrastructure installed for gas production, a major focus of this paper is the potential for CCS in the Marcellus Shale. In considering CCS in relation to the Marcellus Shale, it is important to take into account both the positive and negative impacts of shale gas production. The entire chain of activities involved, not just the injection of carbon dioxide (CO_2), must be examined. In fact, a double chain is involved with the first chain consisting of drilling the gas well, fracturing the formation, producing methane, and pipelining it to market (which could be a power plant). The second chain involves capturing CO_2 from the flue gas of a power plant, transporting it by pipeline to a site in the Marcellus Shale, and ending with injection. This can be represented schematically as shown in Figure 1:

There are two options for injection: 1) CO_2 could be injected into the depleted Marcellus formation, with the Hamilton and Mahantango shale formations acting as the overlying seals, or 2) CO_2 could be injected into a saline formation below the Marcellus Shale; the Marcellus would act as the primary seal (provided it had not been fractured), and the Hamilton and Mahantango acting as secondary seals.

The CO_2 and CH_4 pipelines could use the same right-of-way. This whole chain needs to be evaluated when considering whether CO_2 storage at a Marcellus Shale site is potentially viable.

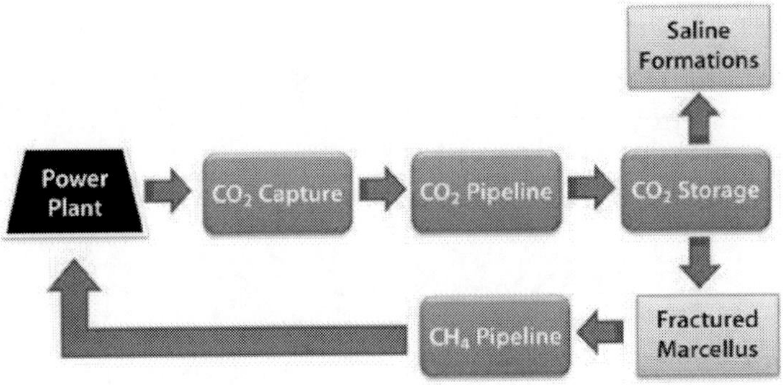

Figure 1. Chain of Activities involved in CCS in relation to the Marcellus Shale.

Scope

Between 2007 and 2009, domestic natural gas production surged by nearly 15 percent as improvements in two technologies—hydraulic fracturing and horizontal drilling—made it possible to produce shale gas at lower cost than other gas in the supply mix. This increased exploitation of shale gas has opened the possibility of CO_2 storage in these shale formations. This report identifies a broad spectrum of interrelated issues that must be addressed in a coherent manner if one wishes to pursue a viable path towards CO_2 storage in conjunction with shale gas production, particularly in the Marcellus Shale.

2.0. Marcellus Shale Basic Geology – Location and Extent

The Appalachian Basin, approximately 300 miles wide and 600 miles long, encompasses a broad area between the Canadian Shield to the north, the Allegheny front to the east, and the Cincinnati arch to the west (Figure 2) (UTBEG, 2008). The basin represents part of an ancient foreland basin in the Eastern United States that contains complex geology formed by a series of continental plate collisions that resulted in the formation of the Appalachian Mountains and large areas of stretched, faulted, and deformed ridges and valleys (USGS, 1993; UTBEG, 2008). The elongate, asymmetrical northeast-southwest trending central axis of the Appalachian Basin is underlain by a succession of sedimentary rocks greater than 10,000 feet (ft) thick (USGS,

1993; UTBEG, 2008). Several of these strata are porous and permeable sandstones that are potential CO_2 geologic storage targets (Milici, 1996). The central axis of the basin is in the vicinity of Pittsburgh, Morgantown, and Cleveland, areas that have large concentrations of CO_2-producing power plants and other large stationary sources of CO_2 (Figure 2) (UTBEG, 2008).

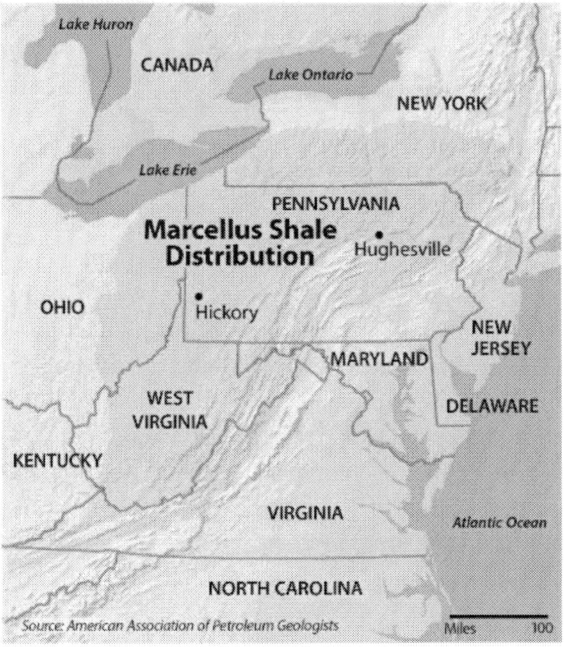

Figure 2. Distribution of the Marcellus Shale Formation.

The black Devonian Shales of the Appalachian Basin, including the carbonaceous Marcellus Shale, have a history of low production and long well life (Harper, 2008). The Middle Devonian Marcellus is highly organic black shale interbedded with medium-gray silty shale and limestone nodules or beds of dark gray-to-black limestone (USGS, 1993). In Pennsylvania, Maryland, and West Virginia the Marcellus Shale contains the Purcell Limestone. The Purcell is composed of gray silty shale interbedded with siltstone, and contains limestone nodules (USGS, 1993).

The Marcellus is the most expansive shale gas play in the U.S., extending on a northeast-southwest trend from west central New York into Pennsylvania, Maryland, Ohio, Virginia, and West Virginia, and covering an area of 95,000 square miles (Figure 2) (Arthur et al., 2008; USDOE, 2009). When the

Hamilton Group is undivided, the Marcellus is classified as its basal unit, underlying the Mahantango Formation of the Hamilton in Maryland, New York, and Pennsylvania. In West Virginia, the contact between the Marcellus and the overlying Mahantango brown shales consists of occasional sandstone beds and concretions, or the Marcellus may lie directly below the Harrel Formation due to a disconformity. This disconformity represents a gap in the geologic record due to either non-deposition or erosion. At its western extent, in eastern Ohio, the Marcellus lies disconformably beneath the Rhinestreet Shale Member of the West Falls Formation (PEP, 2010). The Marcellus Shale typically overlies the Onondaga Limestone; this contact may be sharp, gradational, or erosional (Anderson et al., 1984).

2.1. Depth, Thickness, and Gas Production Potential

The Marcellus, with an average net thickness in the range of 50 to 100 feet, thins to the north, the west, and the south, and pinches out in eastern Ohio, western West Virginia, and southwestern Virginia. The Marcellus reaches subsurface depths of over 9,000 ft along the preserved basin axis; it outcrops to the east and north and subcrops to the west and south. The estimated production depth between 4,000 to 8,500 feet, together with an average gas content of 60 scf/ton, results in a gas-in-place estimate of up to 1,500 tcf (USGS, 1993; USDOE, 2009). Figure 3 shows the stratigraphic column for the Marcellus Shale.

2.2. Stratigraphic Units above the Marcellus Shale

Hamilton Group and Mahantango Formation: Conformably overlying the Marcellus Shale is the Hamilton Group to the north and the Mahantango Formation to the south. The Hamilton Group is a dark gray, brown, or green fossiliferous laminated shale and siltstone at the base, which grades into light gray shale and mudstone with some very fine-grained sandstone in the upper section (Martin, 2006; Milici and Swezey, 2006). To the south and west the Hamilton, with an average thickness of 1,000 feet, grades into the silty shales of the Mahantango Formation (Woodrow et al., 1988). The Mahantango Formation is a dark gray, brown, or green fossiliferous siltstone and shale with coarsening upward trends. The thickness of the Mahantango ranges from zero to 1,000 ft (Butts and Edmundson 1966; Berg et al., 1980).

Tully Limestone: The Tully Limestone is a dark-gray to black silty fossiliferous limestone that overlies the Hamilton Group. It extends from southwest New York to southeastern Ohio and into north-central West Virginia and has a maximum thickness of 200 feet (USGS, 1993; Milici and Swezey, 2006). Several unconformities developed in the Tully and zones of pyrite nodules and pyritized fossils have developed where it thins, such as western New York (Woodrow et al., 1988).

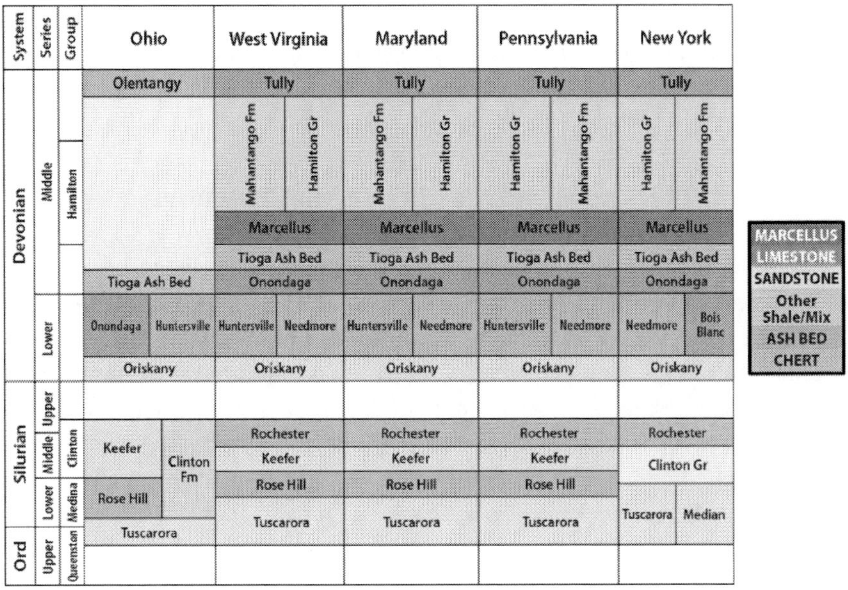

Figure 3. Stratigraphic Column for the Marcellus Shale.

2.3. Stratigraphic Units below the Marcellus Shale

Tioga Ash Bed

The two foot thick Tioga Ash Bed, a regional stratigraphic marker bed, consists of several thin, discrete, volcanic ash falls and often is included as the basal unit of the Marcellus Shale. The Tioga is a gray, brown, black, or green bed with crystal tuff or tuffaceous shale, thinly laminated, with mica flakes (PEP, 2010). Within the central and northern parts of the Appalachian Basin, the Tioga Ash Bed's basal beds are within the uppermost bed of the Onondaga Limestone. In much of the northern Appalachian Basin the Tioga Ash Bed is the uppermost bed lying within the lowermost part of the Marcellus Shale (Hasson and Dennison, 1988; de Witt et al., 1993; USGS, 1993).

Onondaga Limestone

The widely distributed Onondaga Limestone is a gray or gray-blue crystalline limestone that underlies the Tioga Ash Bed/Marcellus Shale. The Onondaga contains scattered chert nodules and has an average thickness of 200 feet (Hall, 1839; Brett et al., 1994). The contact between the Onondaga and the Tioga/Marcellus may be sharp or gradational (Anderson et al., 1984).

Huntersville Chert and Needmore Shale

Underlying the Onondaga Limestone is the Needmore Shale in the east and the Huntersville Chert to the west (Faill et al., 1989). The Needmore Shale is a green-gray to black shale and dark, thin-bedded fossiliferous argillaceous limestone with a maximum thickness of 200 feet. The Needmore grades laterally westward into the Huntersville Chert in western Pennsylvania and West Virginia (Faill, 1997). The Huntersville is highly silicified black shale with brecciated beds that have been recemented with amorphous silica (Woodward, 1943). The Huntersville also contains white fossiliferous chert, glaucaonitic sandstone, and green-gray siltstone (Bartlett and Webb, 1971). The Huntersville Chert, with an average thickness of 200 ft, is considered an important hydrocarbon reservoir in Pennsylvania, West Virginia, and Maryland (USGS, 1993).

Oriskany Sandstone

The Lower Devonian Oriskany is typically a white to light-gray, fossiliferous quartzarenite cemented with locally-variable amounts of quartz or calcite. It can be traced continuously through New York, Pennsylvania, Ohio, Maryland, West Virginia, Virginia, and Kentucky (Diecchio, 1985; Bruner and Smosna, 2008). The Oriskany typically unconformably overlies strata of the Helderberg Limestone or equivalents, and is overlain by Onondaga Limestone, Huntersville Chert, or Needmore Shale, which vary from limestone to chert to shale and are locally sandy (Diecchio, 1985). The Oriskany is a potential geologic storage reservoir, with an estimated capacity of 7,800 million metric tons CO_2 (MRCSP USDOE, 2007).

Rochester Shale

The Silurian Rochester Shale is a dark gray to black calcareous mudstone and shale that underlies the Lockport Dolomite to the west and the McKenzie

Formation to the east. The thickness of the Rochester reaches 65 feet in western New York (Brett et al., 1995). The Rochester Shale could serve as a seal to the underlying Keefer Sandstone and the Tuscarora Sandstone in the event that either one becomes a CO_2 sink.

Keefer Sandstone

The Silurian Keefer Sandstone underlies the Rochester Shale and overlies the Rose Hill Shale. The Keefer has a maximum thickness of over 300 ft in Virginia and thins to the north and southwest (Lampiris, 1976). The Keefer consists of a lower fossilferous iron ore sequence and an upper resistant sandstone unit. The fossiliferous iron ore sequence is gray-red to gray-brown, coarse to very-coarse grained thin bedded calcarenite, and rich in fossils and hematite. The upper unit is white, fine-grained, and thin- to thick-bedded orthoquartzite that is calcareous to the west (Fail et al., 1989). The Keefer is currently being evaluated as a possible CCS target (MRCSP USDOE, 2007).

Rose Hill Shale

The Rose Hill Shale underlies the Keefer Sandstone and conformably overlies the Tuscarora Sandstone. The Rose Hill is 250 to 550 feet thick and consists of fossiliferous dark gray-green or gray-red shale, interbedded with thin fine- to coarse-grained, poorly-sorted, argillaceous sandstone, with a few beds of limestone (Swartz, 1923). The contact with the Keefer Sandstone can be gradational, changing from interbedded shale and calcareous siltstone to fossiliferous hematitic limestone and sandstone (Faill et al., 1989). The Rose Hill Shale and the Rochester Shale may combine to serve as a seal to the underlying Tuscarora Sandstone if it becomes a CCS target.

Tuscarora Sandstone

The Late Ordovician to Early Silurian Tuscarora Sandstone of the Valley and Ridge province of the Appalachian Basin correlates to the Medina Sandstone of the Appalachian Plateau (Faill et al., 1989). The white to light-brown or gray-green Tuscarora is composed of quartzite, quartzarenite, and minor amounts of shale and siltstone (Darton and Taft, 1896; Faill et al., 1989). The Tuscarora, currently being evaluated as a potential CCS target, has an estimated storage capacity of 28,200 million metric tons CO_2 (MRCSP USDOE, 2007).

2.4. Potential Impact on CCS Storage

The continuing development of the Marcellus Shale, one of the largest natural gas plays in the United States, has the potential to both negatively and positively impact the future of CCS in the Appalachian Basin. Although the development of the Marcellus is in the early stages, horizontal well drilling and hydraulic fracturing are key components to production success. Currently, three geologic units are being studied for possible traditional pore space storage of CO_2: (1) the Lower Devonian Oriskany Sandstone; (2) the Middle Silurian Keefer Sandstone; and (3) the Upper Ordovician to Lower Silurian Tuscarora Sandstone (MRCSP DOE, 2007). Due to its low permeability, the Marcellus would serve as one of the seals for these underlying units to effectively prevent vertical migration of injected CO_2. The hydraulic fracturing of the Marcellus has the potential to negatively affect the integrity of this low permeability seal. However, hydraulic fracturing is designed to be confined to a small vertical interval of the Marcellus well above the underlying Oriskany. A well fractured into the Oriskany would be quickly plugged and abandoned due to large water production.

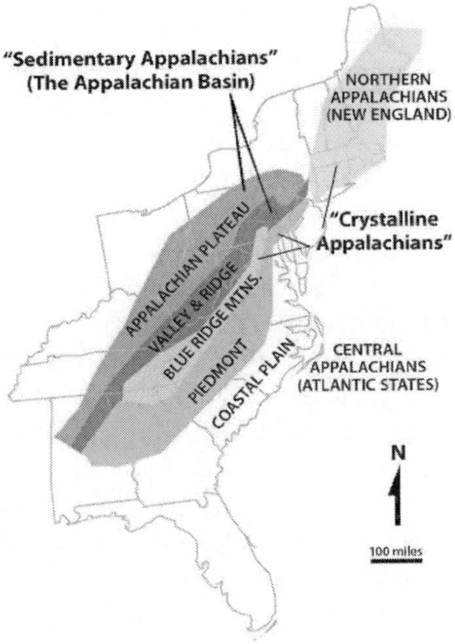

Figure 4. The Appalachian Basin of the United States. From http://3d3dparks.wr.usgs.gov/nyc/valleyandridge/sedimentaryapp.htm.

Previous studies (Nuttall et al., 2005a) have successfully demonstrated that organic-rich shale, like the Marcellus formation, may be favorable to adsorb CO_2. There is potential to safely and effectively store CO_2 in the Marcellus formation (and other similar organic shale formations) due to this phenomenon. In addition, a very significant thickness of overlying low-permeability shale and limestone will remain as an effective seal (e.g., the Hamilton Group). However, fracture stimulation could decrease the integrity of the Marcellus which, in turn, could affect containment for storing large volumes of CO_2 in the underlying Oriskany, Keefer, and Tuscarora sandstone units. On the positive side, fracture stimulation and production of shale gas from the Marcellus could create a new class of potential geologic CO_2 storage targets through adsorption trapping on organic material and clay coupled to a value-added process of enhanced gas recovery (See Section 4).

3.0. Extraction Techniques

The first commercial shale gas well was drilled in 1821 to a total depth of 27 feet into the Devonian Dunkirk Shale. The gas was used by residents of Fredonia, New York to illuminate their homes. Organically rich gas shale reservoirs were once ignored, but are now the focus of increased drilling activity. The challenge to producing economic quantities of natural gas from shale is to release the gas from rock with very small pores and, as a result, very low permeability. Recent advances in drilling and completions (e.g., horizontal drilling, perforating, and hydraulic fracturing), along with higher gas prices, are making shale gas production economical.

Typical pore spaces in shales are not usually large enough for even small molecules like methane to flow at a rate that would make production economical (Figure 5). Consequently, commercial-scale gas production in shales often requires fracturing to provide adequate permeability for gas extraction. Shales may contain some natural fractures, caused by pressure from overlying rock and the natural movement of tectonic plates, that enable some gas flow; and shale gas has long been produced from shales with natural fractures. Recently, however, there has been significant development of gas shales through techniques that create artificial fractures around well bores (fracture stimulation or fracturing).

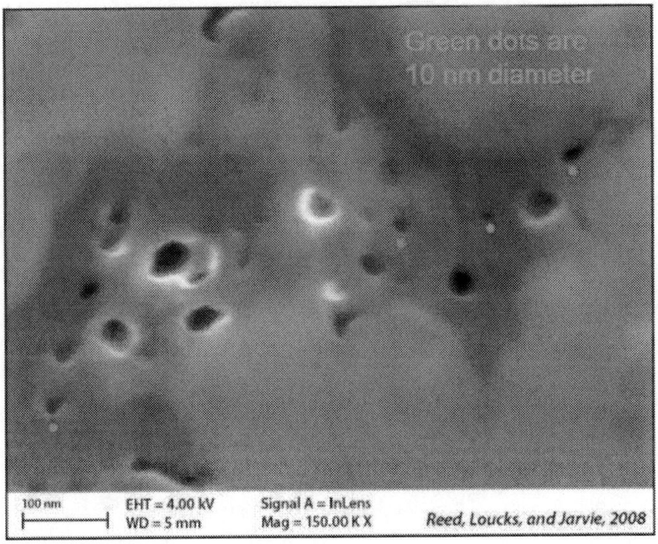

Figure 5. Secondary electron image of nanopores in the Barnett Shale which are so small (20 nanometers) that they impact the passage of methane molecules. Figure attributed to Reed et al., (2008) from Jarvie (2009).

The natural fractures (also known as "joints") in the Marcellus Shale are typically vertically oriented (Figure 6). A vertically-drilled borehole will likely only intersect a few natural fractures, making it difficult to extract shale gas at adequate rates laterally across the formation (Sumi, 2008).

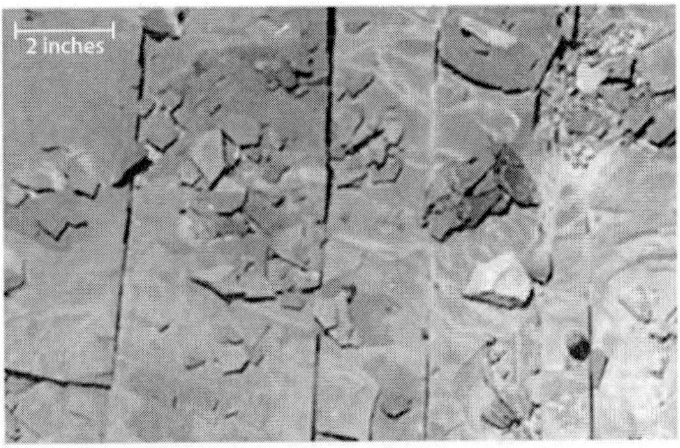

Figure 6. Natural fractures "joints" in Devonian-age shale, typical of fractures in Marcellus Shale. Image from www.geology.

Most oil and gas reservoirs are much more extensive in their horizontal (areal) dimensions than in their vertical (thickness) dimension. Horizontal wells are used in several gas shale formations to enhance lateral gas extraction (Arthur et al., 2008). Horizontal wells are initially drilled vertically. Then, at some distance above the intended target formation (Marcellus Shale) depth (depending on the radius of curvature), the well begins to curve to achieve a horizontal direction that extends through the target formation laterally (Figure 7). As a result, the wellbore in the shale is perpendicular to the most common fracture orientation, which allows intersects with a greater number of fractures (Geology.com, 2010). High yield wells in the Marcellus Shale have been developed using horizontal drilling and fracture stimulation techniques. Some horizontal wells in the Marcellus Shale have initial flows of millions of cubic feet of gas per day, making them some of the most productive gas wells in the eastern United States. However, the long-term production rates of these wells are not currently known (Sumi, 2008).

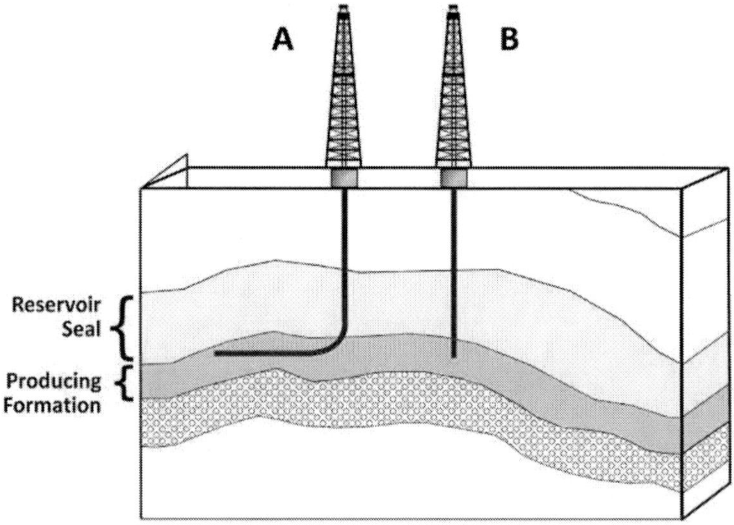

Figure 7. Greater length of producing formation exposure to wellbore in a horizontal well (A) vs. a vertical well (B). Source: EIA, 1993.

A wide range of factors influence the choice between drilling a vertical or a horizontal well to produce natural gas. While vertical wells may require less capital investment on a per-well basis, production is typically at a lower rate, which could affect profitability. For the Marcellus Shale, a vertical well may be exposed to as little as 50 feet of the gas shale, while a horizontal well may be developed with a lateral wellbore extending 2,000 to 6,000 feet within the 50 to 300 feet thick organic-rich shale (Figure 7).

Horizontal wells have a much reduced aggregate surface "footprint" and subsequent surface disturbance resulting from well pads, roads, and pipelines when compared to the equivalent number of vertical wells. More vertical wells (and the associated surface "footprints") are typically needed to extract the same amount of gas as a given horizontal well in the Marcellus (EIA, 1993). Furthermore, several horizontal wells can be placed on multi-well pads for a less intrusive impact to the surrounding area; the decrease in area can also reduce the impacts from noise, traffic, and result in visual changes to the landscape (Arthur et al., 2008). A vertical well can typically cost as much as $800,000 (excluding pad and infrastructure) compared to a horizontal well that can cost in the range of $2.5 million or more (excluding pad and infrastructure) (Arthur et al., 2008).

3.1. Well Development/Stimulation

As far back as the 1980s, horizontal wells were considered viable options and even drilled in Devonian shale units such as the Marcellus. While most of the wells were technical successes based on drilling and final placement, they were ultimately considered failures due to noncommercial gas production rates. The primary reason for insufficient gas production was that most wells required some form of fracture stimulation, which at the time seemed too costly.

Recent growth of horizontal wells in gas shale (for example, in the Barnett shale) is the result of improvements in fracture stimulation technology. Multistage fracture stimulation treatments are now performed on both horizontal and vertical wells to produce hydraulic fractures around the borehole. According to Schlumberger (an oil and gas service company), for almost all gas shale wells, the rock around the wellbore must be stimulated through hydraulic fracturing before a well can produce significant amounts of gas (Sumi, 2008). Well stimulation is typically accomplished through hydraulic fracturing of the target formation. After the well is drilled, cased, and cemented to protect groundwater and prevent the escape of natural gas or other fluids, drillers seal off the interval to be fractured and pump large quantities of water mixed with sand and trace chemicals to modify fluid properties into the shale formation under extremely high pressure to fracture the shale around the wellbore. These induced fractures increase the flow of natural gas through the formation to the wellbore. The sand injected with the fracture fluid acts as a propant that prevents the factures from closing once the pressure is reduced.

The amount of water typically required for hydraulic fracturing ranges from approximately one million gallons for a vertical well to approximately five million gallons for a horizontal well (PADEP, 2010). The recovered fracture water must be either reused in other wells or sent offsite to a treatment facility. To protect surface water resources, a number of states in areas of active shale gas development have become concerned with ensuring proper treatment and disposal of fracture water. For example, in West Virginia, the State Senate passed Bill No. 658 amending the Code of West Virginia (Section §22-11-7C) regarding establishing requirements for use of water resources in Marcellus gas well operations. This bill mandates special permit and reporting requirements for Marcellus Shale gas wells using water resources for fracturing or stimulating gas production. Important reporting requirements under Bill No. 658 include the source and amount of water used, water transport method, and the assurance of proper disposal. The ability to economically stimulate the formation along horizontal well bores has made these wells commercial successes. Wells that were previously drilled vertically to access Marcellus or other formations can be reused to drill horizontally through the Marcellus Shale.

3.2. Well Spacing and Placement

Operators developing the Marcellus Shale are currently using both horizontal and vertical wells to extract the natural gas present in the shale. To effectively manage the resource, the low natural permeability of shale requires vertical wells to be developed at closer spacing than for conventional gas reservoirs. Arthur et al., (2008) have estimated that the spacing for vertical wells in the Marcellus start at approximately 40 acres, while future horizontal wells are predicted to be spaced at intervals closer to 160 acres. By applying these predicted well spacings to a standard section (one square mile - 640 acres) of land, 16 vertical wells would be needed; whereas, the same square mile of resource could be produced from as few as 4-6 horizontal wells drilled from a single multi-well pad.

If development in the Marcellus follows trends established in the Barnett Shale, drilling of longer laterals, bigger fracture jobs with more stages, and more infill drilling can be expected, (As a field matures, additional wells—infill wells, or wells between other wells—may be drilled to increase recovery.) In the Barnett, infills are being drilled down to 10 acres, while refracturing of the first horizontal wells from 2003 and 2004 has commenced. Infills and refractures are expected to improve estimated ultimate recovery from 11 percent to 18 percent (Halliburton, 2008). In addition, multilaterals to reduce the number of pads that need to be developed, especially in urban areas, and water recycling are growing trends that reduce operating costs and minimize potential environmental damage.

4.0. Technical Feasibility for Application of CCS Technology

Aside from its economic and fuel resource benefits, the Marcellus Shale offers two possibilities related to geologic storage of CO_2: (1) the Marcellus could act as a sealing (caprock) formation for injection of CO_2 into limestone and sandstone formations below the Marcellus, and (2) the Marcellus itself could act as a storage reservoir for captured anthropogenic CO_2. The Marcellus Formation is regionally extensive, covering large parts of Maryland, New York, Ohio, Pennsylvania, Virginia, West Virginia, and parts of Canada. It is also relatively thick (ranging from 40 to 900 feet in thickness), with low porosity (~ 0.09) and permeability (5.9 – 19.6 μd). These characteristics make it an excellent caprock formation for CO_2 injected into deeper formations (Soeder, 1988), provided borehole penetrations and fracturing have not comprimised its integrity. On the other hand, the Marcellus itself is also a possible geologic storage option. The Marcellus is a black, organic-rich shale, which could provide favorable CO_2 adsorption as well as available pore space from formation stimulation/fracturing. In this scenario, additional layers above the Marcellus would act as seals.

4.1. Marcellus as a Geologic Storage Target Formation

Continuous, low-permeability, fractured, organic-rich gas shale units are widespread (Figure 9 in Section 5) and are possible geologic storage targets. As part of a project funded by NETL, drill cuttings and cores from the Upper Devonian organic-rich shale units across Kentucky, West Virginia, and

Indiana were sampled, and adsorption isotherms obtained. Sidewall core samples were analyzed for their potential CO_2 uptake and resulting methane displacement. Digital well logs were used to model total organic carbon (TOC) and CO_2 adsorption capacity (Nuttall et al., 2005a, b). Results indicated CO_2 adsorption capacities at 400 psi ranged from a low of 14 scf/ton in less organic-rich zones to more than 136 scf/ton in the more organic-rich zones. There is a direct linear correlation between measured TOC content and the adsorptive capacity of the shale with CO_2 adsorption capacity increasing with increasing organic carbon content.

In shale, natural gas occurs in the intergranular and fracture porosity and is adsorbed on clay and the surface of organic particles (kerogen). This is analogous to methane recovery in coal beds, where CO_2 is preferentially adsorbed and displaces methane. Organic-rich shale may similarly desorb methane in the presence of CO_2. As a result, enhanced natural gas recovery (EGR) may be possible as stored CO_2 displaces methane in gas shale reservoirs.

This preliminary study mentioned above indicates that organic-rich gas shale can serve as a target for geologic storage of significant volumes of anthropogenic CO_2. Highly carbonaceous black shale is the most likely storage reservoir, and the surrounding shale may serve to seal the reservoir. CO_2 trapping by adsorption in gas shale has a higher probability for containment compared to typical pore filled reservoirs (i.e., the gas is bound to the shale). Initial estimates, based on these data, indicate a geologic storage capacity of as much as 28 billion metric tons in the deeper and thicker parts of the related Devonian Ohio Shale in Kentucky. The potential storage resource has not been computed for the Marcellus Shale, but given the greater depths and higher organic content, it could be extremely large. Organic rich gas shale may prove to be a viable geologic sink for CO_2. The extensive occurrence of organic-rich gas shale in Paleozoic basins across North America would make them an attractive regional target for economic CO_2 storage and enhanced natural gas production.

Technical and economic challenges to CO_2 geologic storage and enhanced gas recovery from shale gas reservoirs include (1) potential reduction of the permeability of already low-permeability shale due to differential swelling, similar to that of coal beds, and (2) the potential negative impact on long-term natural gas production, due to CO_2 contamination of produced methane.

Based on data for the adsorption of CO_2 onto organic shales of 14 scf/ton shale to 136 scf/ton shale at 400 psi and the following Marcellus Formation characteristics,

Density = 159 lb/ft^3 of shale Area = 95,000 mi^2
Average Thickness = 100 ft
CO_2 Density$_{Gas}$ = 5.8x10^{-5} scf/ton

the procedure by Nuttall et al. (2005a) can be used to estimate the CO_2 storage potential across the entire Marcellus Formation, which, as a whole, has the potential to store from 17 to 166 billion tons of CO2.

4.2. Potential Risk Associated with Existing Wells

As with any developed or developing gas or oil field, existing wellbores will be a concern that needs to be addressed. The placement, type, and number of well penetrations can have implications for future geologic storage (GS) opportunities in regions overlying the Marcellus. The Marcellus Shale has potential to serve as a regionally extensive caprock formation based on its extent, thickness, and low porosity and permeability (Soeder, 1988). While these wells provide the opportunity to acquire natural fuel resources, they might create potential leakage conduits for CO_2 injected into either deeper formations (like the Oriskany Sandstone) or into the vacant fractured spaces in the Marcellus Formation, possibly creating avenues for CO_2 to migrate to shallower formations or underground sources of drinking water (USDW). However, this potential concern could be modeled, identified, and mitigated with risk assessment software currently available.

Federal and state regulations under the U.S. EPA Underground Injection Control (UIC) program are in place to mandate the construction, operation, permitting, and closure of injection wells in order to protect drinking water sources. This includes the newly proposed UIC Class VI injection well specific to geologic storage of CO_2. Future CO_2 injection wells under Class VI, including into or below the Marcellus Formation, will first require the use of detailed computational modeling to define a calculated region surrounding the well (also called the Area of Review (AOR)) that may be impacted by project activity. The proposed rule for Class VI requires identification and evaluation of all artificial penetrations (including existing extraction wells) and other features that may promote the upward migration of fluids. In response, project operations must plug and/or remediate wells within the AOR as appropriate to prevent fluid migration.

Table 1 outlines gas extraction wells in the state of West Virginia permitted and completed between 2003 and 2009; 2,840 wells were reported by the West Virginia Geological and Economic Survey between 2003 and 2009. Of those 2,840 wells, those intended for, or drilled and completed below, the Marcellus total 2,275 gas extraction wells. As indicated by Table 1, target formations at depths below the Marcellus include the Onondaga, the Oriskany, the Helderberg, and the Tuscarora. In Pennsylvania, by comparison, a total of 2,918 well permits issued for gas extraction from the Marcellus Shale were filed from January – July 2009 (DRNR, 2010).

Table 1. Summary of gas extraction wells and target formations reported in the state of West Virginia from 2003 to 2009. Data provided by the West Virginia Geologic and Economic Survey between 2003 and 2009

Formation	Formation Type	Top of Formation Depth (feet)	Number of Wells
Big Lime/Injun	Shale/Sandstone/ Limestone	4,000 – 6,000	93
Huron/Rhinestreet	Shale	5,000 – 7,000	53
Hamilton Group	Shale	6,000 – 7,000	100
Marcellus	Shale	3,500 – 8,500	2173
Onondaga	Limestone	4,300 – 8,300	28
Oriskany	Sandstone	4,500 – 9,000	45
Helderberg	Limestone	5,500 – 8,800	21
Tuscarora	Sandstone	6,700 – 12,000	8

The state of West Virginia is approximately 24,077 square miles in area, suggesting that at least one gas extraction well (not considering other types of wells) drilled and completed from 2003 – 2009 penetrates the Marcellus Formation (or deeper) every 10.6 square miles. This calculation does not include existing wells in place prior to 2003 (regardless of depth or function) or the fact that more extraction wells can be expected to be drilled into the Marcellus Formation due to promising gas resources. Existing and proposed well distribution is depicted in Figure 8. From an AOR standpoint, CO_2 plume migration typical of a one million tons/yr commercial-scale injection into a high-porosity formation (e.g., Mt. Simon Sandstone) may range from 1 to 1.9 miles from the injection site following plume stabilization (Leetaru et al., 2009).

Newly proposed U.S. EPA Underground Injection Control (UIC) injection wells must account for all existing penetrations within the calculated AOR, several of which may require immediate action to prevent potential leakage. From a federal standpoint, requirements for plugging and abandoning UIC Class I, II and V wells according to 40 CFR 146.10 (minimum AOR radius 1/4 mile from injection well) suggest that the well shall be plugged with cement in a manner which will not allow the movement of fluids either into or between underground sources of drinking water. Project operators must develop a corrective action plan to address improperly completed or plugged wells within the AOR and submit it to the responsible regulatory agency for approval. Procedures for cement placement for plugging Class I, II, and V wells, as described by 40 CFR 146.10, include the following methods:

- **Balance method** – Displace the plugging fluid with cement slurry that is placed through the drill pipe or tubing into the wellbore. Tubing is then slowly pulled back out of the top of the cement, leaving behind a solid plug with minimal contamination by the plugging fluid.
- **Dump bailer method** – A bailer is lowered into a well via wireline and releases a predetermined amount of cement at a given depth. A bridge plug or cement basket is typically previously placed at the specified depth.
- **Two-plug method** – This method involves a top plug, bottom plug, and latch-down type plug catcher. Tubing with the plug catcher is lowered to the desired depth. The bottom plug, followed by the desired cement slurry volume, is pumped into the pipe. The top plug is placed on top of the cement slurry followed by a plugging fluid. Tubing can then be removed leaving a solid cement plug behind (USEPA, 1983).
- **Alternative methods approved by the regulating agency**

Proposed Class VI well rules require a calculated AOR that is reassessed over the duration of the project. Computational modeling is used to forecast lateral and vertical migration of the CO_2 plume and formation fluids. This type of AOR determination is different from assigning a fixed radius AOR, which is the method typically required by existing UIC regulations. The proposed rules require that all active or abandoned wells within the calculated AOR must be identified, and necessary corrective action must be taken in order to prevent the movement of fluids into or between USDWs. Operators must submit the following information to the necessary regulating agency: a description of each well's type, construction, date drilled, location, depth, record of plugging and/ or completion, as well as any additional information. Under the proposed rules, owners or operators of Class VI wells must perform corrective action on all wells in the AOR that are determined to need corrective action using methods necessary to prevent the movement of fluid into or between USDWs, including use of corrosion resistant materials, where appropriate. Well plugging procedures for proposed UIC Class VI wells (40 CFR 146.92) and abandoned wells within the AOR needing corrective action include:

1) Appropriate test or measure to determine bottomhole reservoir pressure
2) Appropriate testing methods to ensure mechanical integrity as specified in 40 CFR 146.89
3) The type and number of plugs to be used
4) The placement of each plug, including the elevation of the top and bottom of each plug
5) The type, grade and quantity of material to be used in plugging; the material must be compatible with the carbon dioxide stream
6) The method of placement of the plugs

In Pennsylvania, the Marcellus Shale natural gas well permit application process requires disclosure of the well location in proximity to coal seams and distances from surface waters and water supplies. The Department of Environmental Protection (DEP) then reviews the application to determine whether the proposed well might cause environmental impacts, conflict with coal mine operations, or exceed well spacing requirements. Operators must submit reports on well completion, waste management, annual production, and well plugging. Pennsylvania law requires drillers to case and cement Marcellus Shale natural gas wells through all fresh water aquifers before drilling through deeper zones known to contain oil or gas. This casing and cement protects groundwater from the fluids and natural gas that will be contained inside the well, and keeps water from the surface and other geologic strata from mixing with and contaminating groundwater (PADEP, 2010).

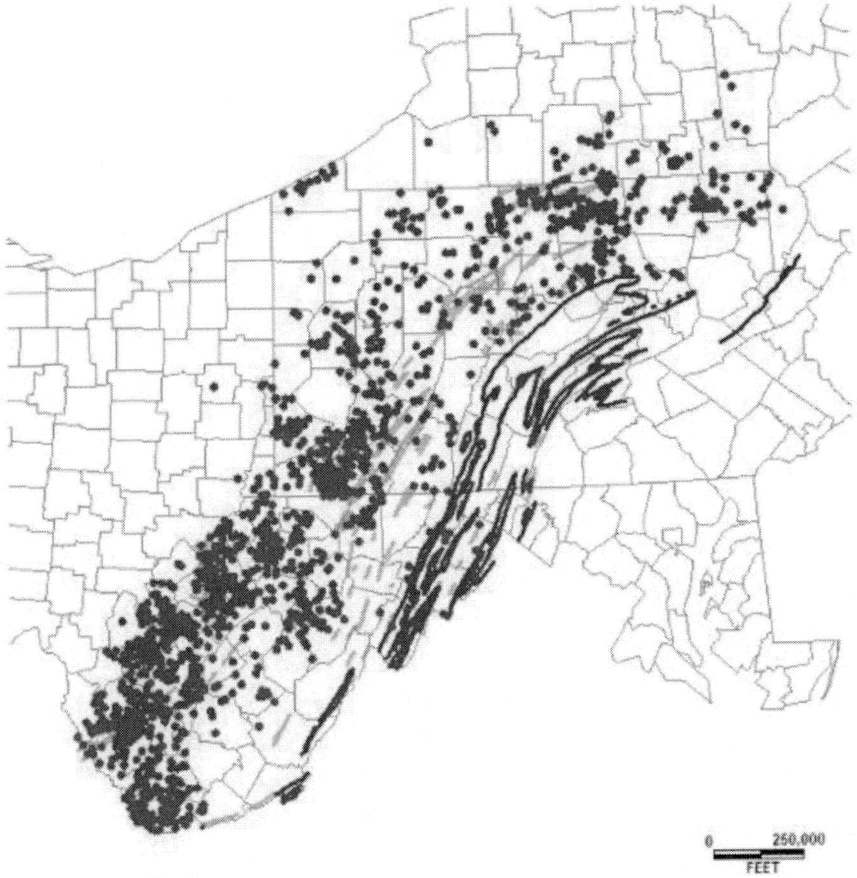

Source: A Shale Tale, Marcellus Odds and Ends, Gregory Wrightstone of Texas Keystone, presented at the 2010 Winter Meeting of the Independent Oil & Gas Association of West Virginia.

Figure 8. Map of the location of Marcellus Shale natural gas wells as of November 2009. Natural gas wells that are drilled into the Marcellus are indicated by red circles, permitted wells are indicated by green circles. The known Marcellus outcrop belt is shown in purple. Know faults that penetrate the Oriskany Sandstone are shown in orange. Due to the map scale (250,000:1) not all wells are represented on the map.

Table 1 provides some insight into the number of new wells drilled in the Marcellus over a six year period simply for extraction purposes. Combining these data with the number of wells already in place and the anticipation of even more Marcellus extraction wells in the near future, suggests two conclusions: (1) for future geological storage projects in or below the Marcellus Formation (in all states where the Marcellus is present), the number of wells in any given calculated AOR assessment could be quite large and will involve a significant level of effort (identification, corrective action plan development and submission, and well plugging); and (2) the number of penetrations and disturbances in the Marcellus Formation (Table 2) (including vertical wells, horizontal wells, and fracturing/stimulation) from both existing wells and future gas extraction efforts could significantly degrade the integrity of the Marcellus as a suitable caprock formation for geologic sequestration for formations below the Marcellus. The loss of the Marcellus as a suitable caprock formation could mean a loss of significant CO_2 storage potential in the northeastern part of the United States. However, even if the Marcellus Shale were to be compromised as a caprock through current drilling activities, the overlying shales of the rest of the Hamilton Group, as well as additional overlying shales, would serve to impede upward migration of sequestered CO_2. These overlying shales would have wellbore penetrations, as they must be drilled to reach the Marcellus, but these would be lower areas of leakage risk than the completed portions of the Marcellus.

Table 2. Estimated number of wells anticipated in Marcellus based on published well spacing data

Marcellus Well Density	
Well Spacing (acres/well)	40 – 160
1 square mile	640 acres
Wells per square mile	4 – 16
Marcellus Area	95,000 mi^2
Total Wells	*380,000 – 1,520,000*

4.3. Summary – Technical Considerations for Geologic Sequestration Near the Marcellus

As summarized in Sections 4.1 and 4.2, the Marcellus Formation provides potentially both (1) an organic shale formation in which CO_2 injection and storage could be favorable due to the shale's affinity to adsorb CO_2, and (2) a caprock formation for geologic storage of CO_2 into the limestone and sandstone formations below the Marcellus. Other general technical considerations include:

- The possibility of combined CO_2 enhanced gas recovery/CO_2 storage in the Marcellus.
- Conversion of "dry wells" into CO_2 injection wells (following UIC Class VI requirements)
- Existing gas extraction well construction materials (cement, casing, etc.) may not maintain integrity in the presence of CO_2 plumes. Well remediation may be required as part of AOR assessment.
- Some of the shale formations above the Marcellus (Hamilton Group and Mahantango Formation could serve as potential caprocks to the Marcellus Formation, should the Marcellus be used for CO_2 storage. While these formations may not be exposed to the horizontal well drilling and hydrofracturing to which the Marcellus is exposed to, due to gas exploitation, they will be penetrated by every well drilled to the Marcellus Formation or below.

5.0. Other Shale Gas Basins

Shale gas reservoir developments are a growing source of natural gas reserves across the United States (Figure 9). Through the use of technology, U.S. gas operators are converting previously uneconomic natural gas resources into proven reserves and increased production. The increase in unconventional gas production illustrates the success of horizontal drilling, fracturing, and completion technologies. This success stems from analysis of geologic and engineering data on decades- to century-old producing areas, identification of unexploited resources, and application of the new drilling and completion technology necessary to convert unconventional resources into reserves.

Across the U.S., numerous geologic basins are recognized as targets for shale gas production; an estimated 35,000 wells were drilled in 2006 (Halliburton, 2008). The Energy Information Administration (EIA) projects that natural gas production from unconventional resources in the United States will increase 35 percent, or 3.2 trillion cubic feet (Tcf), between 2007 and 2030 (EIA, 2009a), with the largest increase expected to come from the development of shale formations in the lower 48 States. Unconventional natural gas production has increased nearly 65 percent since 1998 and has become an increasingly larger portion of total natural gas production, increasing from 28 percent in 1998 to 46 percent of total natural gas production in 2007 (Navigant, 2008). It has been estimated that there are 2,247 trillion cubic feet of gas reserves in the U.S., which is a 118-year supply at 2007 demand levels (Navigant, 2008).

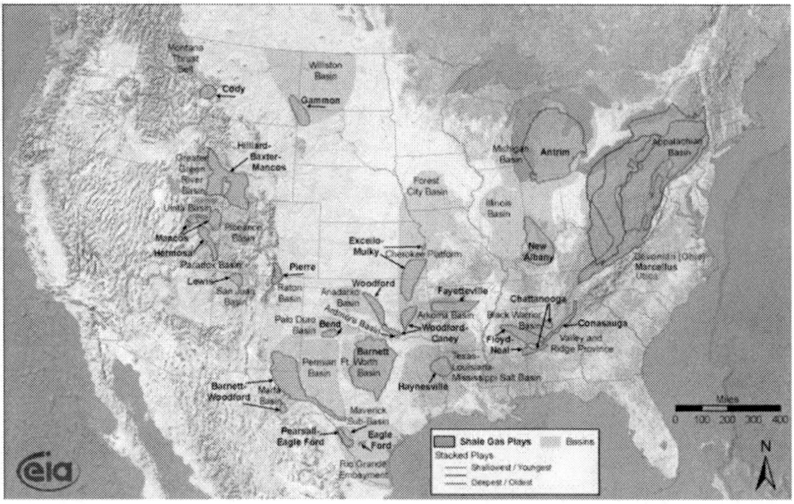

Figure 9. Shale Gas basins in the United States. Map from U.S. Energy Information Administration http://www.eia.doe.gov/oil. Updated May 28, 2009.

Currently, shale gas production is occurring primarily in four fields in the Southwest (Barnett, Woodford, Fayetteville, and Haynesville formations in Texas, Oklahoma, Arkansas, and Louisiana, respectively) and one field in the Northeast (Marcellus, located primarily in New York, Pennsylvania, and West Virginia). In 2008, the gas industry continued to achieve substantial production from these shale formations, leading to a large increase in EIA's estimate of the unproven shale gas resource base to 267 Tcf (EIA, 2009a). The Potential Gas Committee (the PGC consists of volunteer experts who are associated with a wide variety of natural gas industry, governmental, and academic institutions and share a common interest in the Nation's future natural gas supply) reported in 2009 that estimated natural gas resources rose by 35 percent from 2006 to 2008 (PGC, 2009). The current estimated natural gas resource, which includes proven reserves, is largely attributed to shale formations and the ability of advanced drilling and hydraulic fracturing to recover gas. This large increase in reserves arose from reassessments of active and newly developed shale gas plays in the Appalachian basin of the Atlantic area, the Arkoma and Fort Worth basins of the Midcontinent area, several basins of the Gulf Coast area, and the Uinta basin of the Rocky Mountain area.

In the U.S., numerous shale gas basins exist with a resource potential of many hundreds to thousands of Tcf. To date, only a small number of basins have achieved commercial success. In addition to the Marcellus Shale and other shale units in the Appalachian Basin, significant gas shale gas production occurs in the Barnett Shale in the Fort Worth Basin, Lewis Shale in the San Juan Basin, Antrim Shale in the Michigan Basin, Woodford in Oklahoma, Fayetteville in Arkansas, and New Albany Shale in the Illinois Basin (Table 23). In Canada, commercial production has not yet been achieved. To date, potential shale gas plays have been identified in the following regions: Horn River Basin in British Columbia, Montney and Doig in British Columbia, Colorado Group in Alberta and Saskatchewan, and Utica Shale in Quebec.

Some studies suggest that shale gas wells have very steep initial decline rates (Figure 10). Although shale gas wells have a rapid initial decline, a low rate of production is expected to be sustained for decades as old wells provide a solid base of production for 40-50 years (Oil and Gas Investor, 2006).

5.1. Other Selected US Shale Gas Areas

One of the first recognized major shale gas plays, the Barnett Shale in the Fort Worth basin of North Central Texas, is by far the most active shale gas play in the United States (Figure 9). The play encompasses approximately 5,000 square miles in north central Texas. The Barnett Shale of Texas was under investigation as early as 1981, but not until 1995 were the hydraulic fracturing and horizontal drilling technologies available to successfully produce gas at commercial rates. Today two percent of all the gas consumed daily in the U.S. is produced from the Barnett Shale (Haliburton, 2008). It is estimated that production activity in the Barnett Shale may well continue for another 20 to 30 years. The reservoir ranges from 100 ft to more than 1,000 ft in gross thickness and holds from 50 bcf (billion cubic feet) to 200 bcf of gas per square mile.

Producers applied similar approaches to the relatively new Woodford Shale in Oklahoma (Figure 9). Due to its immense development and number of large-scale players, there is an abundance of data on the Barnett shale play; however, less is known about the Woodford and estimates of its potential are still being evaluated. The Woodford play is more faulted (often crossing several faults in a single wellbore) making it easy to drill out of the interval. 3-D seismic and geosteering techniques, in combination with logging "deployed-while-drilling" tools, are important components to successful wells. High silica rocks define the best zones for fracturing, although the Woodford is deeper and has higher fracture gradients than the Barnett. Some producers have obtained promising initial results and the Woodford Shale should continue to expand production.

The Fayetteville Shale is an unconventional play within the Arkoma Basin that covers a large area in northwestern Arkansas (Figure 9). Compared to other shale plays throughout the U.S., the Fayetteville Shale is still an early stage play (commercial production started in 2004) with unique challenges. Productive wells penetrate the Fayetteville at depths between a few hundred to 7,000 feet. This is somewhat shallower than the Barnett. Mediocre production from vertical wells prior to 2007 stalled development in the vertically fractured Fayetteville, and only introduction of long lateral horizontal drilling and hydraulic fracturing has increased initial production rates and drilling activity. At present there is less oilfield infrastructure in place in the area of the Fayetteville, compared to other major plays. 3-D seismic and geosteering are required to ensure that longer laterals of 3,000 or more feet and fracture stages occur in the productive zone. As with most shale gas plays, the growing numbers of wells, need for new infrastructure, and the desire to minimize surface impacts is resulting in an increase in multi-well pad drilling in the Fayetteville Shale.

The Haynesville Shale is a recent addition to the U.S. Shale play. It is roughly located between northern Louisiana, East Texas, and southwestern Arkansas. Its potential upside is its geographical proximity to areas of significant expertise and substantial resources in a play that is thought to be many times larger than the Barnett and with higher gas-in-place. The Haynesville is nearing commercial scale with very promising initial production rates. On the downside, the heterogeneous characteristics of the shale (highly laminated with significant lithologic changes over a few inches) may result in more rapid decline rates than the Barnett or other plays. In addition, at depths of 10,500 to 13,500 feet, this play is deeper than typical shale gas targets, thus increasing costs and technical challenges. Bottomhole temperatures can be 300° F, and wellhead treating pressures can exceed 10,000 psi (Halliburton, 2008). The high-pressure gradient (0.7-0.9 psi/ft) which distinguishes the Haynesville from other shale plays may also result in high decline rates.

Located in the Michigan Basin, the mature Antrim Shale, like the Barnett, has been actively exploited for years and has produced 2.6 Tcf through 2007. Over 9,000 predominately vertical wells have been drilled to relatively shallow depths (500 to 2,000 feet) across a 12 county area of Michigan

(Goodman and Maness, 2008). The gas in the Antrim has a significant CO_2 content (as high as 30 percent in some wells) and wells require de-watering, thus increasing operating costs.

Located in parts of Illinois, Indiana, and Kentucky, the New Albany Shale, has a long history of gas production, but remains in an exploratory stage with gas production occurring primarily in western Indiana and southwest Kentucky. At less than 4,000 feet, the New Albany is much shallower than the Barnett Shale and requires de-watering similar to the Antrim Shale in the Michigan basin. Limited oil field and pipeline transportation infrastructure have hindered development.

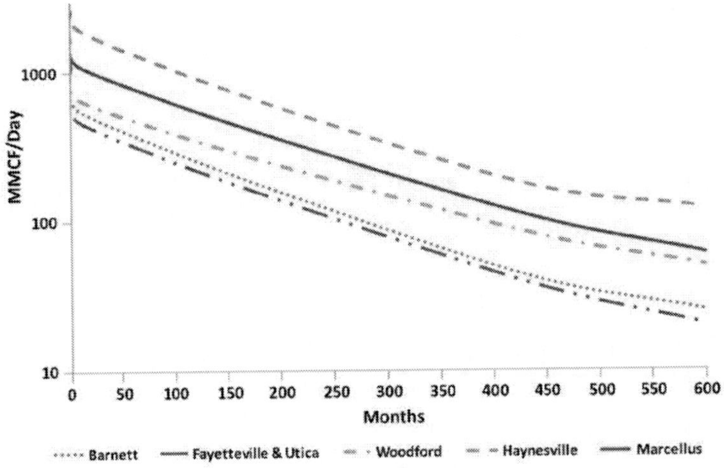

Figure 10. Typical shale gas production decline curves showing rapid initial decline rate followed by multiple decades of low rate production. Adapted and modified from Jarvie, 2009.

Besides the Marcellus Shale, *other Devonian age shale* units are known under different names, each with slightly different geographical boundaries (e.g., Huron, Chattanooga) within the Appalachian basin. The most productive region lies loosely between Kentucky, Virginia, and West Virginia. While estimated reserves are very large, there are operational challenges. Some of the challenges include small operators and service companies with limited resources, a greater interest in coal mining activities, and the extent of the play which tends to be too dispersed. While limited resources hinder development, low transportation costs and proximity to the highly populated East Coast market are attractive aspects of this play.

Table 3. Identified Shale Gas Areas in the United States. Data from various published and unpublished sources

Gas Shale Basin	Marcellus	Antrim	Barnett	Fayetteville	Haynesville	New Albany	Woodford
Estimated Basin Area (mi2)	95,000		5,000	9,000	9,000		11,000
Depth (ft)	4,000-8,500	400-2,000	6,500-8,500	1,000-7,000	10,500-13,500	500-5,000	6,000-11,000
Net Thickness (ft)	50-250	70-120	100-600	20-200	200	100-350	120-220
Total Organic Content (%)	3-12	5.27	4.5	4.0-9.8		7.06	1-14
Total Porosity (%)			4-5	2-8			
Gas Content (scf/ton)		40-100	300-350	60-220		40-80	
Well Spacing (acres/well)	40-160	30-160	60-160		40-560	80	640
Gas-In-Place (tcf)	1,500		327	52	717	86-160	52
Reserves (tcf)	262-500		44	41.6	251	1.9-19.2	11.4
Estimated Gas Production (103cf/day/well)	3,100	20-550	338	530	625-1800	275	415

6.0. Infrastructure Concerns

Pipelines are a mature technology and the most economic method for transporting large quantities of CO_2 over long distances. Gaseous CO_2 is typically compressed to a pressure near 2,200 psi (15.2 MPa) in order to increase its density and avoid two-phase flow regimes, thereby making it possible to pump it as a supercritical fluid and easier and less costly to transport than natural gas. CO_2 can also be transported as a liquid in ships, tank trucks, or in rail tankers within insulated tanks. The first CO_2 pipeline built in the United States is the 225 kilometer Canyon Reef Carriers Pipeline (in West Texas), which began service in 1972 for enhanced oil recovery (EOR) in regional oil fields. Other large CO_2 pipelines constructed since then (mostly in the Western United States) have expanded the CO_2 pipeline network for EOR. These pipelines carry CO_2 from naturally occurring underground reservoirs, natural gas processing facilities, ammonia manufacturing plants, and a large coal gasification project (in North Dakota) to oil fields. Altogether, approximately 5,800 kilometers (3,600 miles) of CO_2 pipelines operate today in the United States (Figure 11).

The North American natural gas pipeline network is frequently mentioned as a model for what future CO_2 pipeline networks might look like, since it interconnects natural gas producing basins with thousands of natural gas distribution companies, power plants, and industrial facilities. The technology, scope, operations, commercial structure, and regulatory framework that characterize natural gas pipelines appear to be useful analogues for a CO_2 pipeline system. Overall, the interstate natural gas pipeline grid consists of approximately 217,000 miles of pipeline with a capacity of about 183 Bcf per day (Figure 11) (EIA, 2009). Between 1998 and 2007, natural gas production in the most rapidly expanding production areas of the Nation (northeast Texas, Wyoming, Colorado, Pennsylvania, West Virginia, and Utah (Figure 13) increased by 129 percent, while proven natural gas reserves grew by 188 percent (EIA 2008; EIA 1999). This increase in production and reserves had a significant impact on the increased number of projects and their location (Figures 11 and 12).

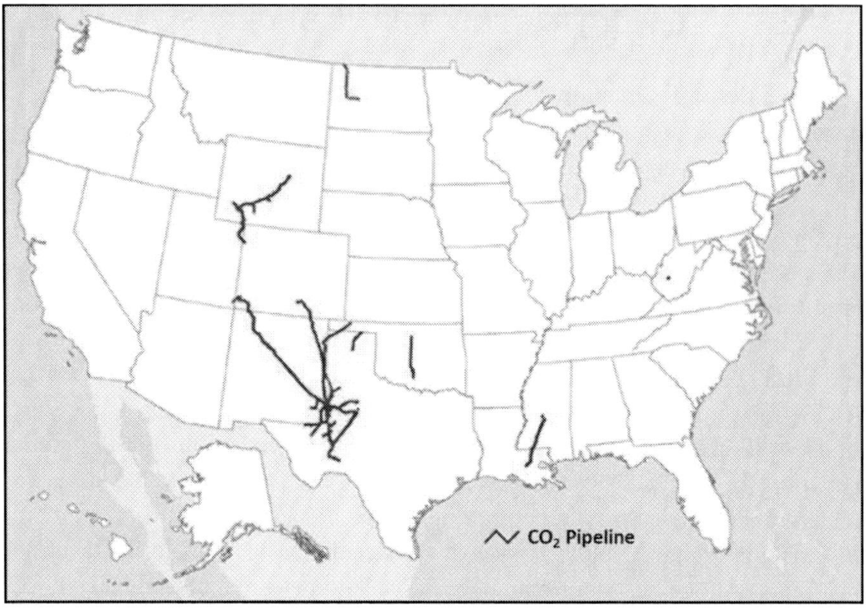

Figure 11. Existing and proposed CO_2 pipelines developed from NatCarb data. (Natcarb is an interactive database developed to show sources of CO_2 and potential sinks for CO_2 storage).

The design of a CO_2 pipeline is similar to that of a natural gas pipeline except that higher pressures must be accommodated, often with thicker pipe. CO_2 pipelines also differ in that they require CO_2-resistant elastomers around valves and other fittings, and their construction includes fracture arrestors every 1,000 feet to reduce fracture propagation, which is more likely in CO_2 pipelines due to their slower decompression characteristics. Another important difference between a CO_2 pipeline and a natural gas pipeline is that the CO_2 pipeline is moving a supercritical fluid that is pumped, not compressed, at booster stations. This is typically done with centrifugal pumps. Inlet pressures at the pumps would be about 1,850 psi (12.8 MPa) and outlet pressures 2,200 psi (15.2 MPa).

In Pennsylvania and West Virginia the development of the Marcellus Shale is spurring the construction of additional natural gas pipeline infrastructure in the region, particularly large-scale expansions of existing pipelines. Much of the region's existing pipeline grid was built to transport gas flows from the Gulf of Mexico. The Marcellus Shale encompasses more than 95,500 square miles and contains significant undeveloped resources, necessitating a reorientation of the region's existing pipeline grid, as well as new gathering pipelines. For example, El Paso Corporation's Tennessee Gas Pipeline Company plans to construct approximately 125 miles of 30-inch pipeline and add approximately 46,000 horsepower of compression facilities in its existing pipeline corridor in Pennsylvania to transport growing Appalachian production to Northeast markets. The project will add capacity of 200 million cf per day (EIA, 2009b).

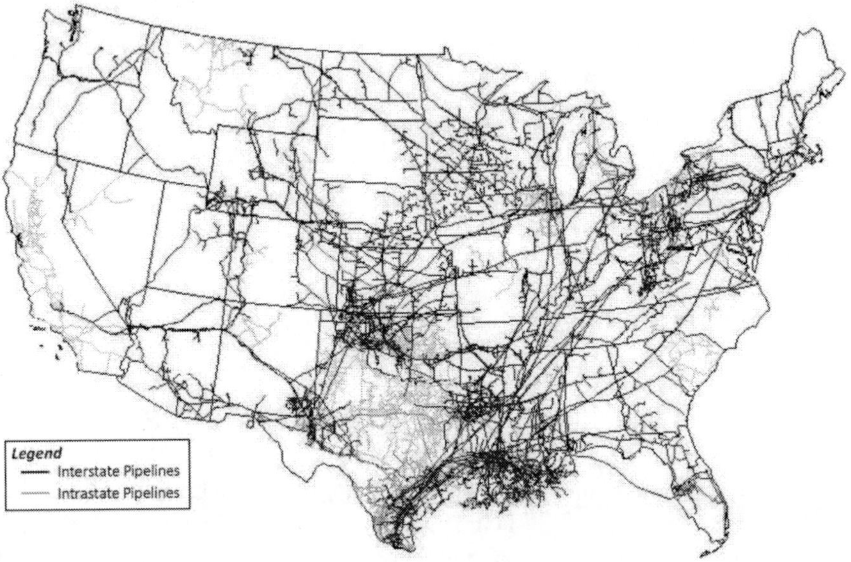

Source: Energy information Administration, Office of Oil & Gas, Natural Gas Division, Gas Transportation information System.

Figure 12. Major recent expansion projects of the U.S. natural gas pipeline network. Image from EIA, 2009b.

Existing CO_2 pipeline systems are largely confined to the Southwest U.S. or to relatively short, dedicated pipelines between CO_2 sources (e.g., ammonia manufacturing plants and gas processing facilities) and nearby CO_2 injection sites (Figure 11). In order to develop a significant CO_2 transportation network on the scale of the existing natural gas pipeline infrastructure, it should be recognized that the investment will require significant capital and may entail competition for the same rights-of-way, material, and manpower resources as that of the natural gas and oil pipelines.

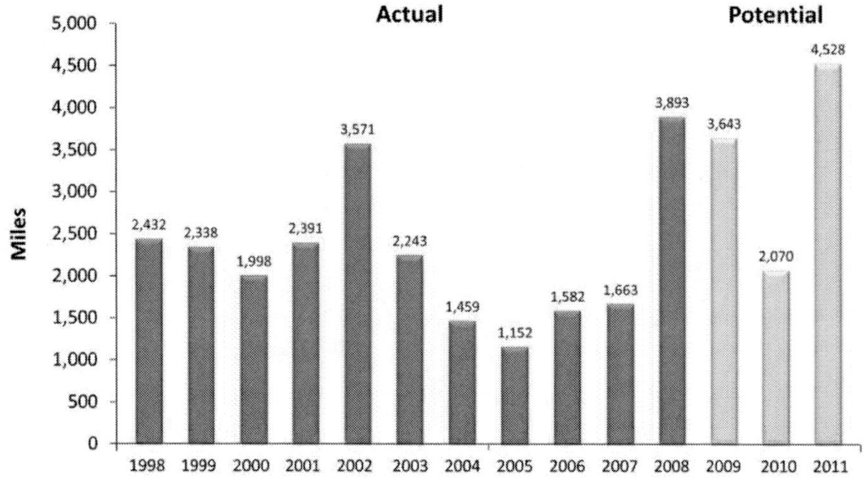

Figure 13. Additions to natural gas pipeline mileage from 1998 to 2011. Image from EIA, 2009b.

7.0. CONCLUSION AND RECOMMENDATIONS

The Marcellus Shale is an extensive formation underlying significant portions of New York, Ohio, Pennsylvania, and West Virginia. As such, it is natural to consider it as a potential sink for CO_2 storage. There are three possibilities for this: (1) storage in the Marcellus Shale itself, (2) storage in formations below the Marcellus, with the Marcellus serving as caprock, and (3) simultaneous gas production and CO_2 storage, similar to enhanced coalbed methane (ECBM) recovery in coal. Because of the very low permeability of the Marcellus Shale, the first of these options is possible only in conjunction with gas production where the formation has been hydraulically fractured to create increased porosity where the CO_2 could be stored. The potential for this option is limited for several reasons: (1) Marcellus Shale wells tend to produce over long periods of time, so it could be a long time before wells may be available and (2) the fracturing process could create fissures which could allow CO_2 to potentially migrate out of the Marcellus. However, low permeability formations above the Marcellus would serve as caprocks to prevent upward movement of CO_2.

The second option of storing CO_2 in formations below the Marcellus may have potential, but a major concern is that the hydraulic fracturing used to recover gas may render the Marcellus unsuitable as a caprock by allowing stored CO_2 to migrate upward through the formation. However, the additional shale sequences above the Marcellus should impede the upward migration of CO_2. Data are needed to evaluate this CO_2 storage option.

The third option entails combined CO_2 storage/methane production similar to CO_2 injection into coal seams, where injected CO_2 adsorbs on organic surfaces and displaces methane. However, no data are currently available to evaluate this option. Furthermore, this option might result in too high a CO_2 content in the natural gas.

The potential for using abandoned gas pipelines for CO_2 transport needs further study because of the different pressure levels, pumping requirements, and corrosion effects of CO_2 compared to methane. Some old gas wells may be useable for CO_2 injection if the cements used can resist CO_2.

As gas production from the Marcellus Shale has increased, this may present an opportunity for CCS in the Marcellus, but more data are needed. As discussed above, this presents an opportunity for CCS in the Marcellus, but more data are needed. There is only one study for CH_4/CO_2 adsorption in shale; and this study indicates that five molecules of CO_2 are adsorbed per molecule of CH_4 produced (compared to three to one for a typical coal). However, the rate of CO_2 adsorption in shale is only one tenth that for coal. Much more data must be generated to allow evaluation of this option, such as the effect of CO_2 on the Marcellus Formation rocks, how much methane can be displaced by CO_2, and adsorption isotherms for methane and CO_2. After the necessary data are generated and further analyzed, it is likely that some option for CO_2 storage can be further analyzed and potentially developed.

REFERENCES

Anderson, R.J., Avery, K.L., et al., 1984, American Association of Petroleum Geologists Correlation Chart; Series 1984.

Arthur, D., Bohm, B., and Layne, M. (2008). Hydraulic Fracturing Considerations for Natural Gas Wells of the Marcellus Shale. Presented at The Groundwater Protection Council 2008 Annual Forum. Cincinnati, Ohio. September 21-24, 2008.

Bartlett, C.S., Jr., and Webb, H.W., 1971, Geology of the Bristol and Wallace quadrangles, Virginia: Virginia Division of Mineral Resources Report of Investigations, no. 25, p. 93. (incl. geologic map, scale 1:24,000).

Berg, T.M., Edmunds, W.E., Geyer, A.R. et al., compilers, 1980, Geologic Map of Pennsylvania: Pennsylvania Geologic Survey, Map 1, scale 1:250,000.

Brett, C.E., and Ver Straeten, C.A., 1994, Stratigraphy and facies relationships of the Eifelian Onondaga Limestone (Middle Devonian) in western and central New York State, *IN* Brett, C.E., and Scatterday, James, eds., Field trip guidebook: New York State Geological Association Guidebook, 66th Annual Meeting, Rochester, NY, no. 66, p. 221-321.

Brett, C.E., Tepper, D.H., Goodman, W.M., LoDuca, S.T., and Eckert, Bea-Yeh, 1995, Revised stratigraphy and correlations of the Niagaran Provincial Series (Medina, Clinton, and Lockport Groups) in the type area of western New York: U.S. Geological Survey Bulletin, 2086, 66 p., Prepared in cooperation with the U.S. Environmental Protection Agency and the University of Rochester, Department of Earth and Environmental Science.

Bruner, K., and Smosna, D., 2008, A trip through the Paleozoic of the Central Appalachian basin with emphasis on the Oriskany Sandstone, Middle Devonian shales, and Tuscarora Sandstone: Dominion Exploration and Production, INC.

Butts, C., and Edmundson, R.S., 1966, Geology and mineral resources of Frederick County, Virginia Division of Mineral Resources Bulletin 80, p. 142.

Darton, N.H., and Taff, J.A., 1896, Description of the Piedmont sheet [West Virginia-Maryland]: U.S. Geological Survey Geologic Atlas of the United States, Piedmont folio, no. 28, p.6.

de Witt, Wallace, Jr., Roen, J.B., and Wallace, L.G., 1993, Stratigraphy of Devonian black shales and associated rocks in the Appalachian basin, IN Roen, J.B., and Kepferle, R.C., eds., Petroleum geology of the Devonian and Mississippian black shale of eastern North America: U.S. Geological Survey Bulletin, 1909-B, p. B1-B57.

Department of Conservation and Natural Resources (DCNR). 2010. Assessment of Geological Sequestration Potential in Pennsylvania.: http://www.dcnr.state. pdf, Accessed April 2010.

Diecchio, R.J., 1985, Regional controls of gas accumulation in Oriskany sandstone, Central Appalachian basin: The American Association of Petroleum Geologists Bulletin, v. 69, p. 722-732.

Energy Information Administration (EIA). 1993. Drilling Sideways – A Review of Horizontal Well Technology and its Domestic Application.

DOE/EIA-TR-0565.: http://tonto.eia.doe.gov/ftproot/petroleum Accessed April 2010.

Energy Information Administration (EIA), 1999 *U.S. Crude Oil, Natural Gas, and Natural Gas Liquids Reserves Annual Report* 1998, http://www.eia.doe.gov/oil reserves_historical.html, Accessed November 1999.

Energy Information Administration (EIA), 2008 *U.S. Crude Oil, Natural Gas, and Natural Gas Liquids Reserves Annual Report* 2007, http://www.eia.doe.gov/oil reserves_historical.html, Accessed November 2008.

Energy Information Administration (EIA), 2009a Updated Annual Energy Outlook 2009 (Reference Case Service Report), http://www.eia.doe.gov/oiaf/servicerpt/stimulus Accessed April 2009.

Energy Information Administration (EIA), 2009b, Expansion of the U.S. Natural Gas Pipeline Network: Additions in 2008 and Projects through 2011, http://www.eia.doe.gov/pub/oil pipelinenetwork/pipelinenetwork.pdf, Accessed September 2009,

Faill, R.T., Glover, A.D., and Way, J.H., 1989, Geology and mineral resources of the Blandburg, Tipton, Altoona, and Bellwood quadrangles, Blair, Cambria, Clearfield and Centre Counties, Pennsylvania: Pennsylvania Geological Survey Topographic and Geologic Atlas, 4th series, 86, p. 209.

Geologists 14, Canadian Society of Petroleum Geologists, Calgary, AB, Canada, p. 157-177.

Geology.com. 2010. Marcellus Shale – Appalachian Basin Natural Gas Play. Geology.com Articles.: http://geology. Accessed February 2010.

Goodman, W. R. and T. R. Maness, 2008, Michigan's Antrim Gas Shale Play—A Two-Decade Template for Successful Devonian Gas Shale Development, Search and Discovery Article #10158 Posted September 25, 2008, p.70.
http://www.searchanddiscovery.com/documents/2008/08126goodman/ndx_goodman.pdf. Accessed April 2010.

Hall, James, 1839, Third annual report of the fourth geological district of the State of New York: New York Geological Survey Annual Report, no. 3, p. 287-339.

Halliburton, 2008, US shale Gas - An Unconventional resource, Unconventional Challenge, 8p,
http://www.halliburton.com/public/solutions/contents/Shale/related_docs/H063771.pdf, Accessed July, 2008.

Harper, J. 2008, The Marcellus Shale – An Old "New" Gas Reservoir in Pennsylvania. Pennsylvania Geology. v 28, no 1. Spring 2008. Published by the Bureau of Topographic and Geologic Survey, Pennsylvania Department of Conservation and Natural Resources.

Hasson, K.O., and Dennison, J.M., 1988, Devonian shale lithostratigraphy, Central Appalachians, U.S.A, in McMillan, N.J., Embry, A.F., and Glass, D.J., eds., Devonian of the world; proceedings of the Second international symposium on the Devonian System; Volume II, Sedimentation Memoir - Canadian Society of Petroleum.

Jarvie, Daniel, 2009, Characteristics of Economically-Successful Shale Resource Plays, U.S.A., Dallas Geological and Geophysical Society Prospect and Technical Expo, Grapevine, Texas May 5, 2009, http://www.wwgeochem.com/ resources/Jarvie+DGGS+May+5$2C+2009.pdf. Accessed April 2010.

Lampiris, N., 1976, Stratigraphy of the clasitc Silurian rocks of central western Virginia and adjacent West Virginia, p. 206.

Leetaru, H., Frailey, S., Damico, J., Mehnert, E., Birkholzer, J., Zhou, Q., and Jordan, P. 2009. Understanding CO_2 Plume Behavior and Basin-Scale Pressure Changes during Sequestration projects though the use of Reservoir Fluid Modeling. Energy Procedia. Volume 1, Issue 1. February, 2009. p. 1799 – 1806.

Martin, J.P., 2006, The Middle Devonian Hamilton Group Shales in the Northern Appalachian Basin: production and potential: at AAPG Eastern Section Meeting, Buffalo, New York.

Milici, R. C., 1996, Stratigraphic history of the Appalachian Basin, *in* Roen, J. B., and Walker, B. J., eds., The atlas of major Appalachian gas plays: West Virginia Geological and Economic Survey Publication V-25, p. 4–7.

Milici, R.C. and Swezey, C.S. 2006, Assessment of Appalachian Basin Oil and Gas Resources: Devonian Shale–Middle and Upper Paleozoic Total Petroleum System: *Open-File Report Series 2006-1237*. United States Geological Survey. http://pubs.usgs.gov/of/2006/1237/of2006-1237.pdf. Accessed 3 March 2010.

MRCSP USDOE, 2007, Carbon Sequestration Atlas of the United States and Canada: United States Department of Energy, Office of Fossil Energy National Energy Technology Laboratory Report, March 2007: http://www.netl.doe.gov/technologies/carbon Accessed March 2, 2010.

Navigant Consulting, Inc., 2008, *North American Natural Gas Supply Assessment*, Prepared for: American Clean Skies Foundation. http://www.navigantconsulting.com/downloads/knowledge_center/North_American_Natural_Gas_Supply_Assessment.pdf, Accessed July 2008.

Nutall, B. C., J. A. Drahovsal, C. Eble and R. M. Bustin, 2005a, CO_2 Sequestration in Gas Shales of Kentucky, Search and Discovery Article #40171, www.searchanddiscovery.net/documents/2005/nutall/index.htm, *Accessed* September 15, 2005.

Nutall, B. C., J. A. Drahovsal, C. Eble and R. M. Bustin, 2005b, Analysis of Devonia black shales in Kentucky for potential carbon dioxide sequestration and enhanced natural gas production, Final Report, DE-FC26-02NT41442, 120p. http://www.uky.edu/KGS/emsweb/devsh/final_report.pdf, Accessed April 2010.

Oil and Gas Investor, 2006, Shale Gas, 24 p. http://www.oilandgasinvestor.com/pdf/ShaleGas.pdf, Accessed April 2010.

Pennsylvania Department of Environmental Protection (PADEP). 2010. Marcellus Shale Factsheet: 0100-FS-DEP4217.: http://www.elibrary.dep.state. Accessed February 2010.

PEP, Phillips Energy Partners, Marcellus Shale, March 2010, http:// phillipsenergypartners.com/buying-mineral-rightsmarcellus-shale, Accessed March 2, 2010.

Potential Gas Committee (PGC), 2009, Natural Gas Resource Estimates and Biennial Report,
http://www.energyindepth.org/wp-content/uploads/2009/03/potential-gas-committee-reports-unprecedented-increase-in.pdf, Accessed June 18, 2009.

Reed, R. M., R. G. Loucks, D. M. Jarvie, and S. C. Ruppel, Morphology, Distribution, and Genesis of Nanometer-Scale Pores in the Mississippian Barnett Shale, AAPG Search and Discover Article, Abstract for AAPG Annual Convention, San Antonio, Texas, http://www.searchanddiscovery.net/abstracts/html/2008/annual/ abstracts/414697.htm. Accessed April 2010.

Swartz, C.K., 1923, Correlation of the Silurian formations of Maryland with those of other areas, *IN* Swartz, C.K., et al., Silurian: Maryland Geological Survey Systematic Report, p. 183-232.

Soeder, D. 1988. Porosity and Permeability of Eastern Devonian Gas Shale. Society of Petroleum Engineers. Institute of Gas Technology.: http://www.pe.tamu.edu/wattenbarger/public_html/ Selected_papers/--Shale%20Gas/SPE15213.pdf, Accessed April 2010.

Sumi, L. 2008. Shale Gas: Focus on the Marcellus Shale. Oil & Gas Accountability Project. Durango, Colorado.

USDOE, 2009, Modern shale gas development in the United States: a primer: United States Department of Energy, Office of Fossil Energy and National Energy Technology Laboratory Report, April 2009.

USEPA. (1983). Technical Manual: Injection Well Abandonment. U.S. Environmental Protection Agency. Office of Drinking Water.: http://www.epa.gov/safewater/uic/pdfs/Historical/ techguide_uic_tech_manual_abandonmt.pdf, Accessed April 2010.

USGS, 1993, Petroleum geology of the Devonian and Mississippian black shale of eastern North America, Bulletin 1909.

UTBEG, 2008. Bureau of Economic Geology, the University of Texas at Austin, Research Environmental Quality, http://www.beg.utexas.edu/environqlty/co2seq/research.htm, Accessed March 3, 2010.

Woodrow, D.L., Dennison, J.M., Ettensohn, F.R., Sevon, W.T., and Kirchgasser, W.T., 1988, Middle and Upper Devonian stratigraphy and paleography of the central and southern Appalachians and eastern mid-continent,

U.S.A., in McMillan, N.J., et al., eds., Devonian of the world, Part I: Canadian Society of Petroleum Geologists Memoir 14, p. 277-301.

Woodward, H.P., 1943, Devonian system of West Virginia: West Virginia Geological Survey Report, v. 15, p. 655.

In: Marcellus Shale and Shale Gas
Editor: Gabriel L. Navarro
ISBN: 978-1-61470-173-6

Chapter 5

SHALE GAS: APPLY TECHNOLOGY TO SOLVE AMERICA'S ENERGY CHALLENGES*

United States Department of Energy

The presence of natural gas—primarily methane—in the shale layers of sedimentary rock formations that were deposited in ancient seas has been recognized for many years. The difficulty in extracting the gas from these rocks has meant that oil and gas companies have historically chosen to tap the more permeable sandstone or limestone layers which give up their gas more easily.

The United States enjoys a rich complement of natural resources, including substantial quantities of fossil fuels—crude oil, coal, and natural gas. These energy sources have helped to fuel our Nation's growth and development for the past two hundred years.

* This is an edited, reformatted and augmented version of the United States Department of Energy publication.

Shale gas well on a Pennsylvania farm. (Photos courtesy of Range Resources).

But American ingenuity and steady research have led to new ways to extract gas from shales, making hundreds of trillions of cubic feet of gas technically recoverable where they once were not.

New technologies are also being applied to make certain that the process of drilling for this valuable resource minimizes environmental impacts.

Barnett shale well at urban location (Courtesy of Chesapeake Energy).

This resource's availability to the American people could not have come at a better time. The calls for reducing our reliance on foreign energy supplies, for reducing our contribution of carbon dioxide to the atmosphere, and for increasing economic growth and wealth creation, can all be met, at least in part, by the development of shale gas. The U.S. Department of Energy (DOE), through the National Energy Technology Laboratory (NETL), has played a historic role in helping to advance the technology that is making shale gas production possible.

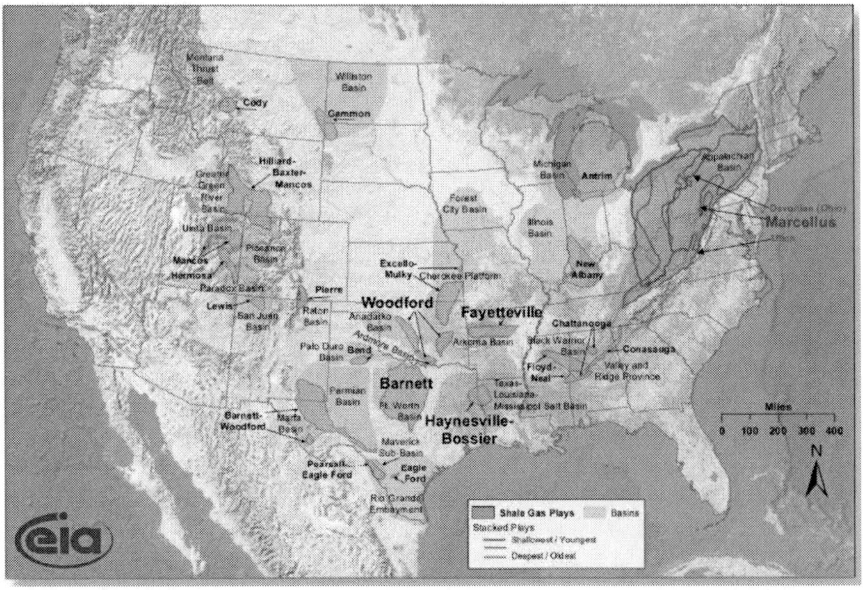

This map, available from the U.S. Energy Information Administration (EIA) at http://www.eia.doe.gov, shows the location and extent of the major shale plays (e.g., Marcellus shale) and the sedimentary basins (regions with thick layers of sedimentary rock containing fossil fuels) where these shale plays are found.

THE RESOURCE

Where Shale Gas Comes from

About 360–415 million years ago, during the Devonian Period of Earth's history, the thick shales from which we are now producing natural gas were being deposited as fine silt and clay particles at the bottom of relatively enclosed bodies of water. At roughly the same time, primitive plants were forming forests on land and the first amphibians were making an appearance. Some of the methane that formed from the organic matter buried with the sediments escaped into sandy rock layers adjacent to the shales, forming conventional accumulations of natural gas which were relatively easy to extract. But some of it remained locked in the tight, low permeability shale layers.

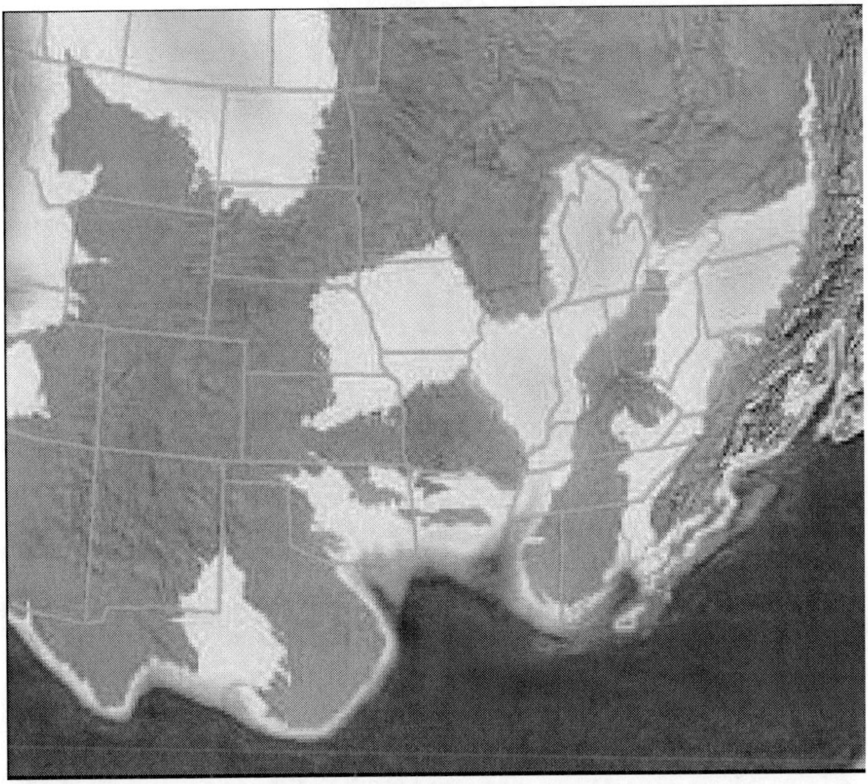

This map of what geologists believe the land looked like 385 million years ago (during the Middle Devonian period) shows the outlines of today's states, and the bodies of water that created the Michigan, Appalachian, and Illinois basins can be seen. (Courtesy Prof. Ron Blakey, Northern Arizona University).

History of Development

The shale gas timeline includes a number of important milestones:

Photo credit Drake Well Museum.

1821 – First U.S. commercial natural gas well in Fredonia, New York, produces gas from shale.

1859 – Edwin Drake demonstrates that oil can be produced in large volumes, launching the U.S. oil industry.

Photo credit Library of Congress.

1860s to 1920s – Natural gas, including gas produced from shallow, low pressure, fractured shales in the Appalachian and Illinois basins, is limited to use in cities close to producing fields.

Photo credit Ohio Historical Society.

1930s – Technology developed to lay large diameter pipelines makes transmission of large volumes of gas from midcontinent and southeastern oil fields to northeastern cities possible; the natural gas industry grows exponentially.

Photo credit Pennwell.

Late 1940s – Hydraulic fracturing first used to stimulate oil and gas wells. The first hydraulic fracturing treatment (not shown here) was pumped in 1947 on a gas well operated by Pan American Petroleum Corporation in Grant County, Kansas.

Early 1970s – Development of downhole motors, a key component of directional drilling technology, accelerates. Directional drilling capabilities continue to advance for the next three decades.

Late 1970s and early 1980s – Fear that U.S. natural gas resources are dwindling prompts federally sponsored research to develop methods to estimate the volume of gas in "unconventional natural gas reservoirs" such as gas shales, tight sandstones and coal seams, and to improve ways to extract the gas from such rocks. Deeper buried shales, such as the Barnett in Texas and Marcellus in Pennsylvania, are known but believed to have essentially zero permeability and thus are not considered economic.

1980s to early 1990s – Mitchell Energy combines larger fracture designs, rigorous reservoir characterization, horizontal drilling, and lower cost approaches to hydraulic fracturing to make the Barnett Shale economic.

2003 to 2004 – Gas production from the Barnett Shale play overtakes the level of shallow shale gas production from historic shale plays like the Appalachian Ohio Shale and Michigan Basin Antrim plays. About 2 billion cubic feet (Bcf) of gas per day are produced from U.S. shales.

Photo credit Pennwell.

2005 to 2010 – Gas production from Barnett Shale grows to about 5 Bcf per day. Development of other major shale plays begins in other major basins.

2010 – The Marcellus shale underlies a significant portion of the mid-Atlantic/NE region—close to East Coast metropolitan natural gas demand centers—and is thought to contain nearly half of the technically recoverable shale gas resource.

Production Trend

Shale gas production continues to increase. In 2009 it amounted to more than 8 Bcf per day, or about 14 % of the total volume of dry natural gas produced in the United States and about 12% of the natural gas consumed in the United States. Production from the Barnett Shale has leveled off, but volumes of gas from the Marcellus, Haynesville, Fayetteville and Woodford shales are growing as more wells are drilled in these plays and as other emerging plays are developed. The EIA projects that the shale gas share of U.S. natural gas production will continue to grow, reaching 45% of the total volume of gas produced in the United States by 2035.

Core from organic Devonian shale formation.

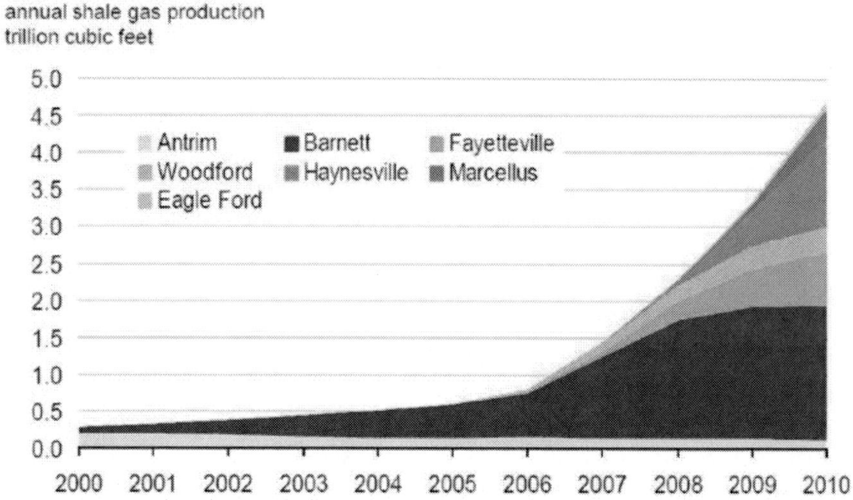

Source: EIA, Lippman Consulting (2010 estimated).

What It Means for Us

The EIA projects that there are 827 trillion cubic feet (Tcf) of natural gas that are recoverable from U.S. shales using currently available technology. The United States currently consumes about 23 Tcf per year, of which we produce about 20 Tcf and import the rest, so the shale gas resource alone represents about 36 years of current consumption. One Tcf of natural gas is enough to heat 15 million homes for 1 year, generate 100 billion kilowatt-hours of electricity, or fuel 12 million natural-gas-fired vehicles for 1 year.

Developing domestic natural gas resources means additional jobs (economic growth) when wells are drilled, pipelines are constructed, and production facilities are built and operated. In addition, higher volumes of available domestic natural gas mean lower fuel or feedstock prices for industries that use natural gas to process or manufacture products. This means fewer jobs lost to lower-cost overseas competitors, as well as lower prices for consumers.

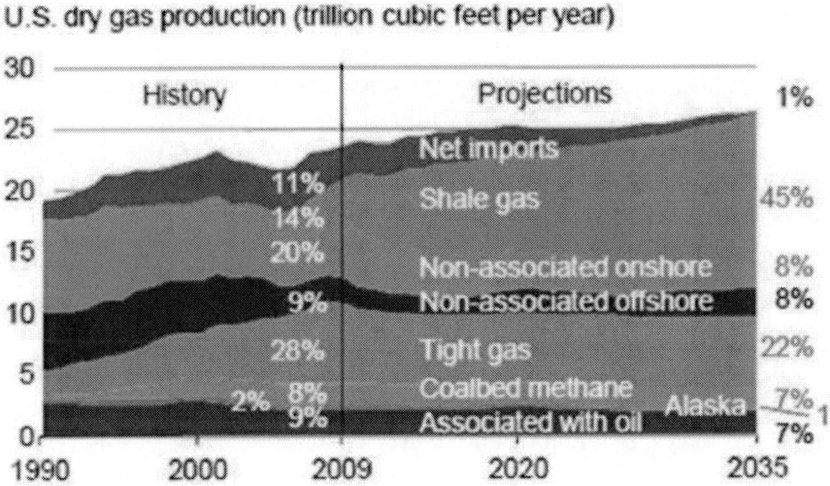

The EIA's Annual Energy Outlook for 2011 shows the contribution of shale gas to U.S. natural gas production reaching 45% by 2035.

Shale gas production also means increased tax and royalty receipts for state and federal government, and increased economic activity in producing areas from royalty and bonus payments to landowners. This influx of revenue can be used to enhance public services.

THE TECHNOLOGY

How It Works

Wells are drilled vertically to intersect the shale formations at depths that typically range from 6,000 to more than 14,000 feet. Above the target depth the well is deviated to achieve a horizontal wellbore within the shale formation, which can be hundreds of feet thick. Wells may be oriented in a direction that is designed to maximize the number of natural fractures present in the shale intersected. These natural fractures can provide pathways for the gas that is present in the rock matrix to flow into the wellbore. Horizontal wellbore sections of 5,000 feet or more may be drilled and lined with metal casing before the well is ready to be hydraulically fractured.

Hydraulic Fracturing

Beginning at the toe of the long horizontal section of the well, segments of the wellbore are isolated, the casing is perforated, and water is pumped under high pressure (thousands of pounds per square inch) through the perforations, cracking the shale and creating one or more fractures that extend out into the surrounding rock. These fractures continue to propagate, for hundreds of feet or more, until the pumping ceases. Sand carried along in the water props open the fracture after pumping stops and the pressure is relieved. The propped fracture is only a fraction of an inch wide, held open by these sand grains. Each of these fracturing stages can involve as much as 10,000 barrels (420,000 gallons) of water with a pound per gallon of sand. Shale wells have as many as 25 fracture stages, meaning that more than 10 million gallons of water may be pumped into a single well during the completion process. A portion of this water is flowed back immediately when the fracturing process is completed, and is reused. Additional volumes return over time as the well is produced.

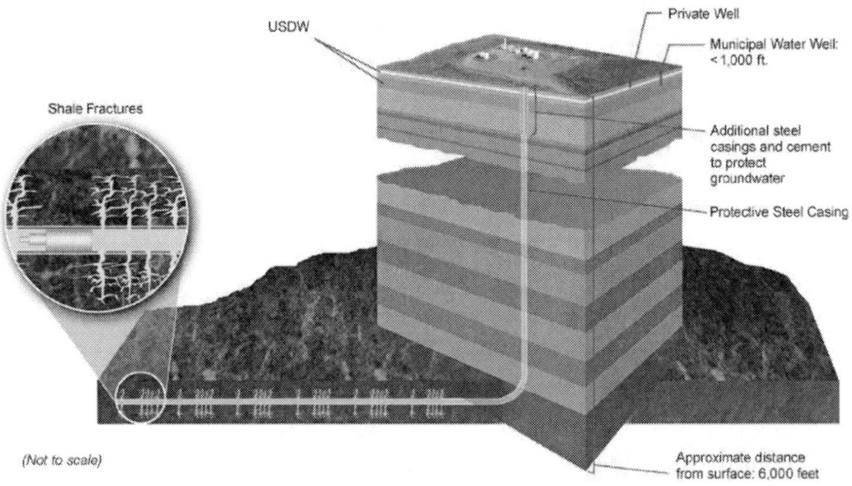

Steel casing lines the well and is cemented in place to prevent any communication up the wellbore as the fracturing job is pumped or the well is produced. Shallow formations holding fresh water that may be useful for farming or public consumption are separated from the fractured shale by thousands of feet of rock.

NETL's Early Contributions

In the 1970s, fears of dwindling domestic natural gas supplies spurred DOE researchers to examine alternative sources of natural gas in unconventional reservoirs such as shales, coal seams, and tight sandstones. NETL helped to advance foam fracturing technology, oriented coring and fractographic analysis, and large-volume hydraulic fracturing. In 1975, a DOE-industry joint venture drilled the first Appalachian Basin directional wells to tap shale gas, and shortly thereafter completed the first horizontal shale well to employ seven individual hydraulically fractured intervals. DOE integrated basic core and geologic data from 35 research wells to prepare the first, publicly available estimates of technically recoverable gas for gas shales in West Virginia, Ohio, and Kentucky.

Hydraulic fracturing job on Marcellus multi-well pad in Pennsylvania.

DOE researchers gathering data from one of a series of cored shale wells in the Appalachian Basin in the early 1980s.

DOE's important contributions to shale gas development have been recognized by many. According to Penn State University's Dr. Terry Engelder, a recognized expert on the Marcellus Shale, DOE's Eastern Gas Shales Research Program "helped expand the limits of gas shale production and increased understanding of production mechanisms. It is one of the great examples of value-added work led by the DOE." In his recent paper summarizing thirty years of gas shale fracturing, George E. King, Global Technology Consultant for Apache Corporation, states that "Technology developments in the North American Devonian shale during the late 1970s and proceeding into the 90s, chiefly from a loose alliance of the U.S. Department of Energy, the Gas Research Institute and numerous operators, combined to collectively produce several breakthroughs … horizontal wells, multi-stage fracturing and slick water fracturing." Fred Julander of Julander Energy, a 36-year independent producer and a member of the National Petroleum Council, has stated that "The Department of Energy was there with research funding when no one else was interested and today we are all reaping the benefits. Early DOE R&D in tight gas sands, gas shales, and coalbed methane helped to catalyze the development of technologies that we are applying todat.

For example, EQT, an independent producer in Pittsburgh, PA, has been developing the Huron Shale in Eastern KY using air drilling technology that relies on electromagnetic telemetry (EMT) to directionally drill horizontal wellbores. EQT reports that it is currently producing more than 100 million cubic feet per day (MMcfd) from its Huron wells and believes the resource potential could be as much as 10 Tcf of gas equivalent. The EMT technology now offered by Sperry Drilling (a Halliburton service line) has its roots in DOE research from the 1980s and 90s. "In the early 1980s, the industry as a whole did not have a clear vision for producing gas from shales and benefited from DOE involvement and funding of EMT technology… there is a clear line of sight between the initial research project and the commercial EMT service available today," states Dan Gleitman, Sr. Director – Intellectual Asset Management, Halliburton.

While decades of technological enhancements stand behind the suite of tools and methodologies that make shale gas production possible, publicly funded R&D has played an important role. NETL continues to manage a suite of research projects focused on increasing the supply of domestic natural gas to the consumer, in an environmentally sustainable and increasingly safe manner.

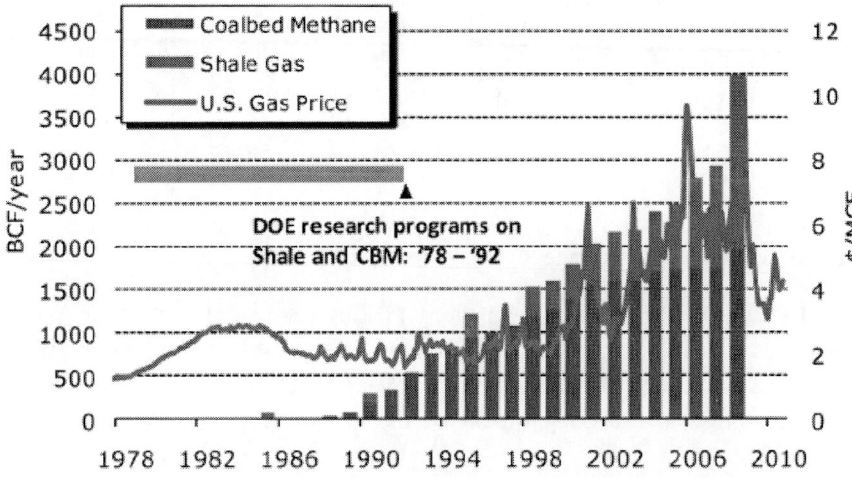

DOE research during the 1980s played a role in the growth of unconventional gas production that is now helping to reduce the price of natural gas to consumers.

What's Next

What DOE Is Doing Now

Currently, NETL is actively involved in advancing technologies that can help producers develop shale gas resources in the most environmentally responsible manner. Research is under way to find improved ways to treat fracture flowback water so that it can be reused or easily disposed of and to reduce the "footprint" of shale gas operations so that there is less disruption of the surface during drilling and completion operations.

DOE is refocusing the work done under Section 999 (Subtitle J) of the Energy Policy Act of 2005 on safety, environmental sustainability, and quantifying the risks of exploration and production activity.

Fracturing make-up water is stored in lined pits to protect groundwater. (Photo courtesy of John Veil, Argonne National Laboratory).

DOE is working closely with the U.S. Environmental Protection Agency (EPA) as it carries out an exhaustive study to quantify the potential risk of hydraulic fracturing to underground sources of drinking water. NETL is also collaborating with the Department of Interior to enhance understanding of these risks.

Recent years have witnessed a number of initiatives to address the challenges of producing shale gas, sponsored by states, environmental groups, industry advocacy groups, and research organizations. DOE is exploring creation of a Shale Gas Initiative, in cooperation with public, private and non-governmental stakeholders, to build on these efforts and identify "best practices" that could be used by both operators and regulatory agencies to raise the bar on safety and environmental sustainability during shale gas development.

Fracturing trucks on location at a Pennsylvania Marcellus location. (Photo courtesy of John Veil, Argonne National Laboratory).

The U.S. Department of State has launched a U.S.-China Shale Gas Resource Initiative to help reduce greenhouse gas emissions, promote energy security and create commercial opportunities for U.S. companies. To date, the effort has engaged hundreds of Chinese technologists, facilitated a Chinese delegation's visit to a U.S. shale gas development operation, and created interest in American unconventional gas technologies through forums and workshops.

DOE has worked with states through the Ground Water Protection Council (GWPC) to develop and maintain the Risk-Based Data Management System (RBDMS). Nationwide, 20 states and one Indian Nation now use the RBDMS to help operators comply with regulations. DOE has recently enhanced the RBDMS to track and record data related to hydraulic fracturing treatments. DOE has also funded in part, a Hydraulic Fracturing Chemical Registry to be hosted by the GWPC and Interstate Oil and Gas Compact Commission (IOGCC). This website will be a means for the industry to voluntarily supply hydraulic fracturing chemical data in a consistent and centralized location.

In 2009, DOE teamed with IOGCC to form a Shale Gas Directors Task Force to serve as a forum for states to share insights on issues and innovations related to shale gas development at the local, state and federal levels. More information is available at www.iogcc.org and http://groundwork.iogcc.org.

While it will be impossible to extract shale gas without some temporary disruption to the rural landscape, new and existing technologies can be employed to limit this disruption, to mitigate any surface impacts, and to minimize impacts to other natural resources in the process.

Well sites require temporary disturbance of the landscape while drilling is underway. (Marcellus well site photo courtesy of Range Resources).

Where to Find out More

You can find out more about shale gas from these resources:

- **NETL website –** The National Energy Technology Laboratory has a complete list of research projects, with details about objectives, accomplishments, expected benefits and results, at http://www.netl.doe.gov/.
- **DOE website –** The Department of Energy has information available on Department objectives and accomplishments related to natural gas at http://energy.gov/energysources/naturalgas.htm.
- **Marcellus Shale Coalition website –** This website has general information provided by an organization "committed to the responsible development of natural gas from the Marcellus Shale geological formation and the enhancement of the region's economy that can be realized by this clean-burning energy source" at http://marcelluscoalition.org/home/.
- **Groundwork –** The IOGCC website focuses on shale gas regulatory information at http://groundwork.iogcc.org.
- **Publications –** A number of publications have been produced by NETL and others that help to explain shale gas and the technologies involved. These include:

– "Modern Shale Gas Development in the United States – A Primer," available for download at http://www.netl.doe.gov/technologies/oil-gas/publications/EPreports/Shale_Gas_Primer_2009.pdf

– NETL's "E & P Focus Newsletter" provides updates on various shale gas research projects, available for download at http://www.netl.doe.gov/technologies/oil-gas/ReferenceShelf/epfocus.html

– "An Emerging Giant: Prospects and Economic Impacts of Developing the Marcellus Shale Natural Gas Play," available for download at http://www.alleghenyconference.org/PDFs/PELMisc/PSUStudyMarcellusShale072409.pdf

– "The Economic Impacts of the Pennsylvania Marcellus Shale Natural Gas Play: An Update," available for download at http://marcelluscoalition.org/wp-content/uploads/2010/05/PA-Marcellus-Updated-Economic-Impacts-5.24.10.3.pdf

– "Developing the Marcellus Shale," available for download at http://www.pecpa.org/sites/pecpa.org/files/downloads/Developing_the_Marcellus_Shale_0.pdf

– "Water Resources and Natural Gas Production from the Marcellus Shale," available for download at http://pubs.usgs.gov/fs/2009/3032/

– "Homegrown Energy: The Facts About Natural Gas Exploration of the Marcellus Shale," available for download at http://www.marcellusfacts.com/pdf/homegrownenergy.pdf
– "The Future of Natural Gas: An Interdisciplinary MIT Study," available for download at http://web.mit.edu/mitei/research/studies/naturalgas.html

INDEX

D

E

I

J

K

L

M

N

O

P

Q

R

S

T

U

V

W

Y